Ambrose Loomis Ranney

Practical suggestions respecting the varieties of electric currents and the uses of electricity in medicine

Ambrose Loomis Ranney

Practical suggestions respecting the varieties of electric currents and the uses of electricity in medicine

ISBN/EAN: 9783337306502

Printed in Europe, USA, Canada, Australia, Japan

Cover: Foto ©berggeist007 / pixelio.de

More available books at **www.hansebooks.com**

PRACTICAL SUGGESTIONS

RESPECTING THE

VARIETIES OF ELECTRIC CURRENTS

AND THE USES OF

ELECTRICITY IN MEDICINE.

WITH HINTS RELATING TO THE SELECTION AND CARE OF ELECTRICAL APPARATUS.

BY

AMBROSE L. RANNEY, M. D.,

PROFESSOR OF THE ANATOMY AND PHYSIOLOGY OF THE NERVOUS SYSTEM IN
THE NEW YORK POST-GRADUATE MEDICAL SCHOOL AND HOSPITAL ; PRO-
FESSOR OF NERVOUS AND MENTAL DISEASES IN THE MEDICAL
DEPARTMENT OF THE UNIVERSITY OF VERMONT ; FELLOW
OF THE NEW YORK ACADEMY OF MEDICINE ; MEMBER
OF THE NEUROLOGICAL SOCIETY OF NEW YORK ;
OF THE NEW YORK COUNTY MEDICAL
SOCIETY, ETC.

NEW YORK:

D. APPLETON AND COMPANY,

1, 3, AND 5 BOND STREET.

1885.

PREFACE.

FROM time to time for several years I have given instruction in electro-therapeutics and electro-diagnosis to post-graduate medical students; and it now constitutes a part of my yearly course before the classes of the Medical Department of the University of Vermont.

While engaged in preparing a forthcoming work upon the diagnosis and treatment of nervous diseases, it occurred to me that this subject would make an interesting and valuable chapter—especially as so much in neurology depends upon a knowledge of electricity and electrical apparatus. I therefore prepared the following pages, with that special intent—as a brief summary of such information as I deemed most essential to a scientific and successful use of electricity in medical practice.

I have been persuaded by my students to anticipate the publication of my proposed volume somewhat through the "New York Medical Journal," and to produce this special chapter in the form of a small guide for their use.

I am partly actuated in so doing by the pressing need of such a guide in my classes; and also by the belief that some members of the profession may thus be reached and possibly benefited who are not particularly interested in neurology, and would probably not buy the larger volume.

The fourteen plates at the end of the book (after Ziemssen and Flower) will aid the reader in treating morbid states of the motor or sensory apparatus.

It is but just to express here my indebtedness to Mr. H. E. Waite for special facilities afforded me by his factory. He has kindly allowed me to use these facilities at will in the practical investigations that I have made respecting the construction of batteries and the actions of electrical currents. Many of the modifications in electrical apparatus which I have devised have been manufactured by him under my personal supervision and guidance.

AMBROSE L. RANNEY, M. D.

NEW YORK CITY.

156 MADISON AVENUE, *November, 1885.*

CONTENTS.

PART I.

ELEOTRO-PHYSICS.

LECTURES ON

ELECTRICITY IN MEDICINE.

In the diagnosis and treatment of nervous diseases, no agent is more generally applicable than electricity. Its brilliant and often instantaneous effects and the prevalent belief among the laity that electricity is practically identical with the vital forces of the human body have conduced largely to the general use and *frequent abuse* of this important agent.

Thousands of electric batteries are sold yearly by the various manufacturers to persons both in and out of the medical profession. Many who buy them are utterly ignorant of the principles of their construction, and equally so of the indications for their use. A very large majority of the medical profession possess only a faradaic battery or a magneto-electric machine. They employ such a battery upon every case which to their mind requires electricity. A few, in our larger cities, own a galvanic battery; but, as a rule, those who do so are unable to repair it themselves when the connections become oxidized or when it fails to act from a multitude of other causes. In my experience, it is very uncommon to meet a medical practitioner (outside of those who are specially interested in neurology) who thoroughly understands electro-physics and many important facts relating to the uses for which special forms of batteries are best

1

adapted. I have deemed it wise, therefore, to include in this course of lectures a terse and practical statement of the more important facts which should be mastered before the treatment of disease by electricity is attempted, and to shed some light upon the forms of current which are indicated in the treatment of many of the nervous diseases commonly encountered. I shall include in these remarks some practical suggestions respecting the selection of batteries and the care of them. The uses of electricity in diagnosis, as well as its therapeutical properties, will be also presented in as concise a form as is comportable with clearness of statement.

Part I.

ELECTRO–PHYSICS.

Under this heading we shall first discuss the varieties of electric currents which may be produced (the faradaic, galvanic, magneto-electric, and static). We shall then consider the construction of a galvanic cell and its many modifications. It is important that you know the principles of construction of the various galvanic cells offered to the profession for medical uses, as well as the advantages and disadvantages of each as a part of a medical outfit. In the third place, you should be made familiar with many new terms which are commonly used to-day in electrical literature, and also the application of Ohm's law to electrical problems. Finally, you should acquire a familiarity with the many attachments to a battery. These are essential to its proper use, and their purposes should be well understood. Under this heading I shall give you some practical hints respecting the selection, care, and repair of an electrical outfit for medical purposes.

VARIETIES OF ELECTRIC CURRENTS.

A few of the more important facts relating to this agent (which we are constantly called upon to employ in the treatment of various types of disease) should be thoroughly understood by all who intend to use it. Time will not permit of a detailed description of the different properties of electric currents. These can be acquired from any of the standard works upon physics. It is necessary, moreover, that such points as are presented here should be briefly and simply stated.

The GALVANIC CURRENT (called also "*voltaism*," the "*battery current*," and the "*constant current*") is one which is derived by chemical decomposition or heat from one or more pairs of elements directly. When the body is placed between two electrodes connected with such a battery in action, the current traverses the part of the body embraced between the electrodes before it returns to the battery—starting at the positive pole (the anode), circulating through animal tissue, and returning to the negative pole (the cathode). The polarity remains unchanged under all circumstances.

Muscular contractions are produced only when the current is closed or broken, or when its intensity is increased. A very weak current fails to produce muscular contractions. In connection with the description of the tests employed in the diagnosis of nervous diseases ("Med. Record," March 22 to June 28, 1884), suggestions have been made by me which may be reviewed in this connection with advantage.

By peculiar arrangements of the elements of a battery, the galvanic current can be modified as follows: (1) To produce heat (cautery battery); (2) to insure chemical changes in living tissues (electrolysis); and (3) to aid in many of the mechanical arts, such as electro-plating, electric lighting, telegraphy, etc.

The FARADAIC CURRENT (called also the "*induced*" or "*interrupted current*") differs from the galvanic in that it is an induced current of high tension, which is produced by the magnetizing and demagnetizing of a bar of soft iron or a bundle of soft-iron wires by means of a galvanic current.

The circuit of the generating cell is made to pass through a coil of insulated wire, known as the "*helix*," which surrounds the iron to be magnetized, *but it does not itself pass to the electrodes and thus to the patient.**

When the current of the generating cell passes through the helix, the soft iron is magnetized and draws the interrupter in contact with it.† This breaks the circuit and de-

---

* This is a point which can not be too strongly impressed upon the minds of the profession. Its accuracy can be readily proved. If the binding-posts of the primary coil of a faradaic machine be united by means of a large copper wire, the current generated in the galvanic cell which runs the interrupter will pass through the wire rather than through the helix which surrounds the iron core (because the wire affords less resistance). The interrupter will then remain stationary, as the iron core is no longer magnetized. Again, the interrupted or faradaic current has no chemical properties. This would not be the case if the current of the generating cell passed to the binding-posts of the faradaic machine. Finally, the external resistance of the human body is far in excess of that afforded by the helix, and this alone would prevent the galvanic current of the generating cell from traversing the animal tissues (the circuit of the greatest resistance).

It is not uncommon for agents of the various manufacturing companies to show a prospective purchaser of a faradaic machine the galvanic cell which works the "interrupter," and to endeavor thus to leave the impression that a galvanic current (as well as the primary and secondary faradaic currents) can be conveyed by a faradaic machine to a patient. Such a statement is untrue, and, if made, indicates either ignorance or dishonesty. Subsequent diagrams will render the mechanism of a faradaic machine intelligible to the reader.

† The expression "in contact" is not strictly correct. The interrupter never actually touches the iron core, because its magnetic action ceases before it reaches it.

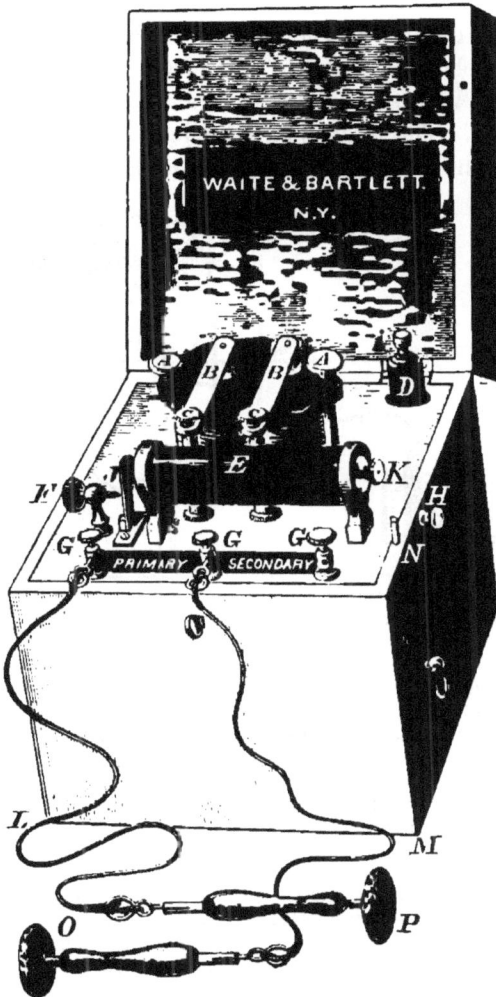

FIG. 1. - ONE OF THE MANY FORMS OF FARADAIC MACHINE. *BB*, connecting rods attached to the elements of the exciting cell ; *D*, a drip-cup, in which the zinc element is placed when not in use (it should contain mercury) ; *E*, primary and secondary coils ; *F*, adjusting screw for the interrupter (*J*) ; *G*, binding-posts ; *K*, plunger ; *L, M*, rheophores ; *O, P*, electrodes. The next figure will explain the action of the different parts. The faradaic instruments of different manufacturers vary more or less in their mechanical devices and perfection of workmanship, but the principle of all is the same.

magnetizes the iron. The interrupter is then returned to
its former place by a spring. This step reconnects the generating cell with the helix, and again allows the iron to be
magnetized. The interrupter is again drawn in contact with
it. Thus the current is constantly broken and restored by

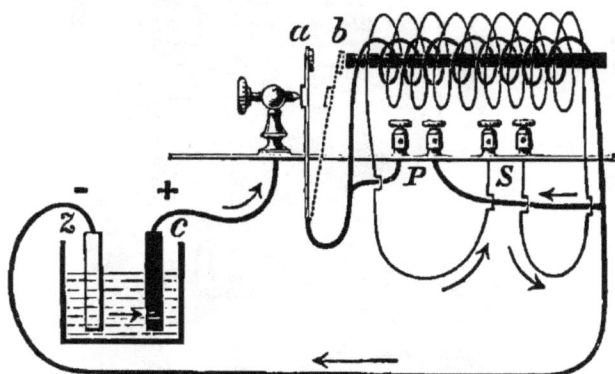

FIG. 2.—A DIAGRAM DESIGNED BY THE AUTHOR TO ILLUSTRATE THE CON-
STRUCTION AND ACTION OF A FARADAIC MACHINE.— Z, zinc element; C,
carbon element; P, binding-posts for the primary coil; S, binding-posts of
the secondary coil; a, the interrupter when the circuit is passing to the
helix; b, the interrupter when the circuit is broken. The screw (shown in
contact with a) allows of the adjustment of the interrupter to the handle of
soft-iron wires within the primary helix, thus making the interruptions fast
or slow at the will of the operator. The patient is connected with the battery in action by means of cords attached to the binding-posts at P or S.
These cords are not shown in the diagram, but are shown in Fig. 1. The
arrows show the direction of the currents. The zinc is marked as the negative element (−), and the carbon as the positive (+) element of the battery.
Note that the wire of the primary coil is represented as coarser than that of
the secondary; that the secondary coil has no connection with the elements
of the cell; that the current going to the primary binding-posts is generated
by the iron core, and is not that which originates in the galvanic cell; and
that the interrupter has a small piece of platinum soldered upon it where it
comes in contact with the screw, so as to prevent oxidation at that point.
Patients feel the current made by the "break" more than that from the
"make" of the circuit; hence one electrode apparently gives a stronger
current.

a simple device known as the "interrupter," or "automatic
circuit-breaker." An *induced current within the iron core* of

the helix is thus produced. This is the current which passes
through the electrodes to the patient.

Much ingenuity has been shown in the construction of
the "interrupter" of a faradaic machine. It is very desirable that *slow and rapid interruptions* may be produced at
the will of the operator. If a machine only insures rapid
interruptions, the slow interruptions can be effected by the
use of an "interrupting electrode."

The power of producing electrolysis, and some other
chemical properties peculiar to the galvanic current, are
wanting in the faradaic.

Never attempt to combine a galvanic and faradaic battery.
Separate cells should be employed for each, as the faradaic
battery requires a cell of greater capacity than those used
in portable galvanic machines. They may be placed in the
same case, but each should be perfectly independent of the
other. If a battery is designed for transportation, it is
best to have one of each rather than two combined in one
case.

The faradaic current is an *alternating current*, i. e., one
which goes in opposite directions at each make and break
of the circuit. It is strongest when the current is broken.
These facts are not generally recognized by the profession.
The polarity changes with each interruption. The so-called
"cathode" of a faradaic battery is felt the strongest by
the patient.

The helix of a faradaic machine is usually surrounded
by a secondary coil of wire, known as the "*secondary helix.*"

This coil has no *connection with the elements of the generating cell*. The current produced within it is induced by
the passage of the current (formed within the generating
cell) through the "primary helix," which magnetizes and
demagnetizes the iron core. It is, therefore, called the
"*secondary current.*" It has high tension, is alternating,

and is employed in telephonic lines, chiefly on account of its high intensity. It is modified in strength by regulating the amount of the secondary coil which overlaps the primary—the smaller the extent of the overlap, the weaker the current. A sliding tube of metal is sometimes made to pass over the primary coil, or between it and the primary coil. This accomplishes the same results as if the helix was movable.

The primary and secondary coils of a faradaic machine are made of wire of different thickness and length. Many of those sold are poorly constructed. They are the most important features of the instrument, and should be made with the greatest care and of the best materials. A fine finish of brass mountings and varnish does not always indicate good workmanship in the coils themselves.

Induced currents develop in the individual coils of the wire forming the primary spiral, as well as in the iron core which it invests, and also in the secondary helix. These currents are of no therapeutical value without the iron core, as they lack sufficient intensity.

If the secondary helix is composed of very fine wire, the current induced within it is extremely painful. The number of coils and the thickness of the wire selected for the primary and secondary helix should be graduated to a proper relation to each other and the electro-motive force of the generating cell employed.

The term *"primary current"* is often used as synonymous with the galvanic. It is incorrectly applied by some authors to that faradaic current which is induced by the magnetizing and demagnetizing of the soft-iron core of the helix.

Static electricity is derived from friction. A revolving plate, or by preference several plates, of glass may be employed as a generator. Static electricity is sometimes called *franklinism,*

This form of current has high tension, but it possesses no chemical properties. When a patient is charged with it, it is necessary that the *chair upon which he sits should be insulated* by glass or rubber under the legs. When highly charged, sparks may be elicited from the body of the patient-through the clothing. This is not always desirable, but it often renders its application easy, and on that account is mentioned as an advantage by manufacturers of these instruments.

Experimentation has shown that this form of electricity is accumulated upon the periphery of the object charged (as explained in all works upon physics). It apparently does not permeate very deeply below the surface.

The MAGNETO-CURRENT (called also the "*dynamic current*") is derived from a *permanent* or *electro-magnet, in front of which an armature is made to revolve.* The armatures are composed of a core of soft iron wound with insulated wire, and the currents produced are formed within them by *breaking the lines of magnetic force.*

The stronger the magnet, the more rapid the breaks made in the current by the revolving armatures, and the greater the number of turns in the spiral wire of the armatures, the more intense is the current.

This form of current possesses great electro-motive force or intensity. Currents of this kind are of an alternating character. By means of an automatic commutator (polarity changer), they may be carried, however, in one direction, and, when so, they assume properties similar to those of galvanic currents. Electric lighting, electro-plating, and many other similar applications to the mechanical arts, are to-day accomplished by means of dynamo-machines at a minimum cost as compared with battery currents. They are practically obsolete as a machine for medical purposes, as the current is unsteady when hand-power is employed.

"Magneto-current" machines (which are turned by a
crank when in use) are often sold to physicians. They are

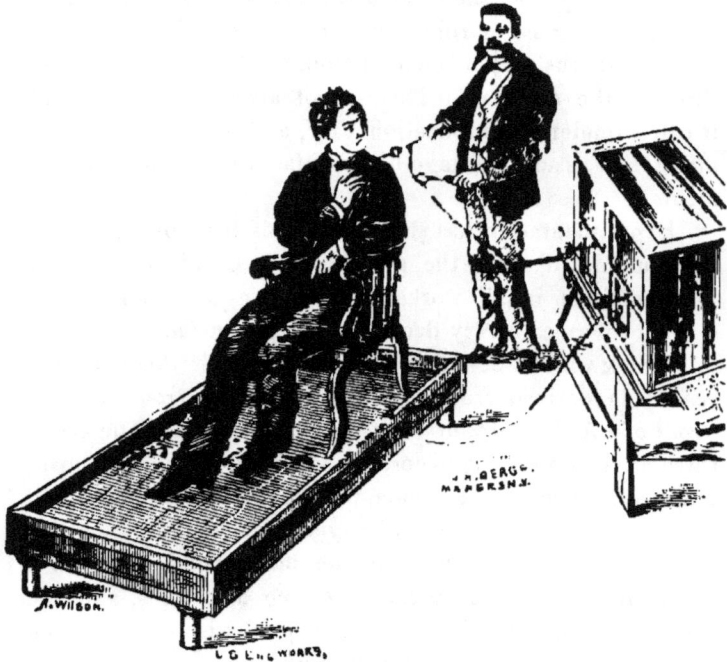

Fig. 3.—A Static Machine in Use. The "direct spark" is here represented as
being drawn, i. e., the patient being charged positively and the electrode being
connected with the negative pole of the machine. No Leyden jars are employed
in this form of administration of static electricity. The "indirect spark" is
more commonly employed than the direct, one pole being attached, in this case, to
the insulated platform, the other being grounded by a brass chain, and the electrode being grounded by a chain attached to a gas-pipe, a water-faucet, or the like
(See Fig. 30.)

of little value. They cost as much as a good faradaic instrument, and are not to be compared with the latter. The
current generated in both is practically the same, but it is
irregular in point of strength in the magneto-current machine, and uniform in the faradaic instrument.

THE GALVANIC CELL—ITS VARIETIES, AND THE GENERAL
PRINCIPLES OF ITS CONSTRUCTION.

All substances have an electrical condition which is inherent or capable of being developed. This condition is known as the " POTENTIAL " of a body. The electrical condition of the earth (which may be regarded as fixed and as a reservoir without limit) is used as a standard of comparison of the " potential " of any given substance.

Those bodies from which electricity tends to flow toward the earth are known as "*positive bodies*" or bodies of "*high potential.*" They are designated by the plus sign (+). Those which tend to draw electricity from the earth are called "*negative bodies*" or bodies of "*low potential,*" and are designated with the negative sign (−). Almost every known substance may, therefore, be classified either as positive or negative under certain circumstances. Subsequent explanations will make this more apparent to you.

When we speak of the " RELATIVE POTENTIAL " of two bodies, we mean the *difference in degree* of the potential of each.

The bodies thus compared must both be positive or negative. One metal, for example, may have a potential seventy times that of the earth, and another one hundred times that of the earth. Both may be positive, yet one is negative as compared with the other. Two such metals have been happily compared to reservoirs at different levels (De Watteville). The tendency between two bodies of different relative potentials is for the current to flow from the body having the highest potential to that possessing the lower potential, thus tending to establish an equilibrium between them. In a galvanic cell, the element most corroded by the fluid of the cell has the highest potential (positive element).

The difference in equilibrium between the "potentials" of two bodies regulates the intensity of what is known as the "ELECTRO-MOTIVE FORCE" of the bodies selected; because the want of equilibrium is the force which starts the flow of an electric current in all cases. The size of the elements has nothing to do with it.

The simplest form of a GALVANIC CELL consists of two bodies (whose potentials differ widely) immersed in a fluid which tends to excite chemical decomposition of one of the elements. Zinc and carbon are commonly selected for the elements and dilute sulphuric acid for the exciting agent.

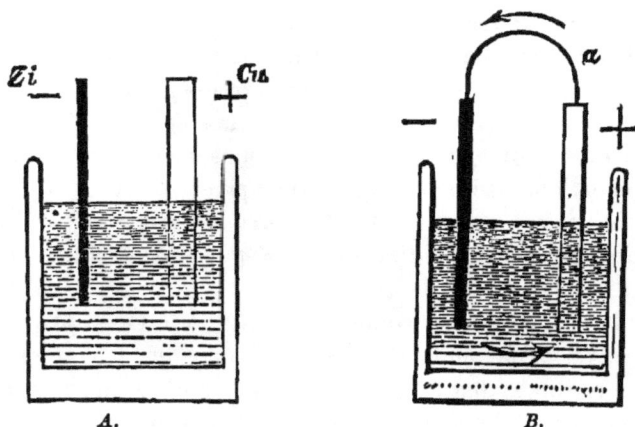

FIG. 4.—A SIMPLE GALVANIC ELEMENT (after Erb). *Zi*, zinc element; *Cu*, carbon or copper element. The fluid is composed of diluted sulphuric acid or a solution of some of the salts. In *A* the circuit is open; in *B* it is closed by a wire connecting the elements. The arrow shows the direction of the current outside of and within the cell.

The zinc is strongly acted upon by the fluid, while the carbon is not; hence the zinc becomes the positive element and the carbon the negative. In most batteries of this type the zinc is covered with mercury (amalgamated), to render the action of the cell more uniform and to prevent local action upon the zinc. It also tends to preserve the zinc.

An apparently discordant fact should be remembered— i. e., that the wire connected with the carbon of such a cell (the negative element of the cell) is the *positive pole* of the battery. This is because the electric current passes through the liquid from the zinc to the carbon, and back through the external circuit from the carbon to the zinc. (Fig. 4, *B*.)

When the elements of a cell are connected externally by a wire, a current of electricity flows continuously from the cell-elements through both the wire and fluid. This is known as a " *complete* " or " *closed circuit.*"

The RESISTANCE offered to the passage of the current from the *carbon to the zinc* is the " external resistance "; that between the *zinc and the carbon* is known as the " internal resistance " of the cell.

The INTERNAL RESISTANCE of a galvanic cell may be modified as follows: (1) By the *distance* between the elements; (2) by the *size of the elements;* (3) by the *intervention of some foreign body* (such as a porous cup) between the elements; and (4) by the *character of the fluid* in which the elements are immersed. The nearer the elements are placed, the larger their size, the more direct the passage of the current, and the better the conducting power of the fluid used, the less the internal resistance of the cell, and *vice versa.* The internal resistance of a cell may vary between a fraction of an ohm and one hundred ohms, according to its construction and its excitants.

The EXTERNAL RESISTANCE is modified by the *length,* the *diameter,* and the *character of the conductor* employed. When any substance (such as the human body, for example) is placed between the electrodes, the resistance offered to the passage of a current by the interpolated substance must be added to that afforded by the conductors themselves. The resistance of the human body varies from 600 to 18,000 ohms. It is extremely low in subjects afflicted with general anasarca—probably because an excess of fluid renders the body a good conductor. The average resistance of the human body is not much above 2,500 to 3,500 ohms. The resistance afforded by the body is modified (1) by the saturation of the electrodes; (2) by the moisture of the surface of the body ; (3) by the tissues through which the current is directed; (4) by pressure made upon the electrode; and (5) by many other factors which will be mentioned hereafter, among which the addition of salt to the water in which the electrodes are moistened is a very important one.* In cautery batteries the external resistance is increased about $\frac{1}{273}$ for every degree centigrade when the temperature of the platinum wire is raised (De Watteville). Thus heat may be a factor in modifying the external resistance to be overcome.

The relative resistance of living tissues is represented by the following figures (100 being taken as the maximum): The eye, 4 ; muscle, 6 ; nerve, 10; cartilage, 20; tendon, 20; fat, 75; bone, 100; skin, 100. Thus the eye offers only $\frac{1}{25}$ the resistance afforded by skin and bone ; muscle, $\frac{1}{16}$; nerves, $\frac{1}{10}$; cartilage and tendon, $\frac{1}{5}$; and fat, $\frac{3}{4}$. The epidermis, when dry, is practically a non-conductor of electrical currents.

Respecting this point, De Watteville happily remarks that " the human body may be compared to a vessel bound

---

* In testing this point lately by means of a Brenner's rheostat, I found the resistance from right palm to left palm, in a boy of thirteen years of age, to be 17,500 ohms, when pure water was used and the electrodes pressed firmly againt the skin of both palms. Adding a teaspoonful of salt to the water and again soaking the sponges reduced the resistance to 7,500 ohms. This illustrates well the necessity for so simple a precaution when employing electric currents upon animal tissues.

with a poorly conducting material (the skin), unequally packed with non-conducting solid particles, the interstices being filled up with a saline fluid of fair conductive power. The parts most densely packed with solid particles are represented by the bones; those where liquid predominates, by the muscles. Between the two are found the nerves, viscera, etc."

Before we leave the discussion of the various forms of electric currents employed in medicine, it may be well to impress upon your minds some of the more important facts relating to faradaism and galvanism by means of a table in which the two are contrasted with each other. Such a table is not, to my knowledge, to be found in any work upon electricity. It may prove of service to you in many ways:

THE FARADAIC CURRENT	THE GALVANIC CURRENT
Is an "induced current." Is produced by the *magnetizing and demagnetizing* of a core of soft iron.	Is due to *chemical decomposition* of one or more of the elements of a galvanic cell.
Its polarity changes with each "make" and "break" of the circuit.	Its polarity is constant. The negative element of the cell becomes the positive pole of the battery.
The current is an interrupted one.	The current is a continuous one.
It produces muscular contractions of an apparently continuous character, provided the interruptions are very rapid.	It does not produce muscular contractions, except when the intensity is increased or when the circuit is made or broken.
The polarity is inconstant, because the currents constantly alternate in their direction.	Each pole has a special therapeutical action peculiar to it under all circumstances.

Is seldom administered by the so-called "polar method."	Is administered chiefly by the "polar method."
Wide separation of the poles intensifies the pain.	Separation of the poles does not materially intensify the pain.
The "secondary current" has greater penetrating power than the "primary current." Neither equals the galvanic current in this respect.	Has a remarkable power of penetrating animal tissues placed "in circuit."
It has no chemical properties. It may be modified by an automatic commutator, so as to throw its currents constantly in the same direction, as in a dynamo-machine. In this case it possesses chemical attributes.	Possesses inherent chemical properties, hence its power of producing electrolysis, and its use in electro-plating, electro-lighting, etc.
A galvanometer will show only one deflection, i. e., the difference in strength of the "make" and "break" currents. This deflection is the same under all circumstances when the machine is in use. It does not, therefore, indicate the strength of the current conveyed to the patient.	Produces galvanometer deflections which are proportionate to the strength of the current employed.
There is no difference in the action of the poles.	The anode is the sedative pole; the cathode is the stimulating pole.
The faradic instrument makes a "buzzing noise" when in action.	A galvanic instrument gives no external manifestation of activity, because it has no interrupter.

ELECTRICAL UNITS.—Before the construction of an electric battery and the modifications in such an apparatus necessary to produce special effects are considered, it is important that you familiarize yourselves with the various units of measurement employed in electricity, and their symbols. These are as follows:

THING MEASURED.	SYMBOL.	NAME EMPLOYED FOR UNIT.
Quantity.	Q.	*Coulumb.*
Current.	C.	*Ampère* or *Weber.*
Electro-motive Force.	E. M. F. or E.	*Volt.* (contraction of Volta).
Resistance.	R.	*Ohm.*
Capacity.	K.	*Farad.* (contraction of Faraday).
Work or Energy.	W.	*Joule.*
Power.	P.	*Watt.*

A COULUMB is the quantity that passes in one second of time against one ohm of resistance under an electro-motive force of one volt. We use this term as we do "pints" or "quarts" in speaking of fluids. One coulumb will decompose 92 microgrammes of water, and thus evolve 10·4 microgrammes of hydrogen.

An AMPÈRE is the current produced by one volt against one ohm of resistance. In medical practice, the milliampère is generally accepted as the unit of current-strength. An ampère will decompose ·00142 of a grain of water.

A VOLT is the electro-motive force necessary to produce a current of one ampère against an ohm of resistance. It practically equals the electro-motive force of one Daniell's cell. We speak of a battery as of so many volts just as we designate an engine as of so many horse-power.

An OHM is the resistance necessary to allow of one ampère of current under an electro-motive force of one volt. It is equivalent to a piece of telegraph-wire one hundred metres in length and of a certain definite sectional area, or

a column of mercury one square millimetre in diameter and
1·05 metre in height.

A FARAD is the capacity of a condenser which would contain a charge of one coulumb under an electro-motive force
of one volt.

A JOULE is the amount of electric energy absorbed when
a coulumb falls one volt. It is equivalent to about $\frac{1}{4}$ of the
heat required to raise one gramme of water at 0.° C one degree, or ·7373 foot-pounds.

A WATT is the power developed by one ampère falling
one volt. It is equivalent to $\frac{1}{726}$ of a horse-power.

The prefixes "*meg*" and "*micro*" denote million and
millionth. For example, a megohm is one million ohms;
and a microhm is a millionth of one ohm.

The names selected for the various units of measurement
are taken from those of prominent electro-scientists (Ohm,
Volta, Faraday, Ampère, and others).

OHM'S LAW.—We are now prepared to consider the law
of electric currents discovered by Ohm, by which the intensity of a current that will result from any combination
of cells may be mathematically computed, and many other
electrical problems solved. It may be thus stated :

$$\text{INTENSITY OF CURRENT} = \frac{\text{ELECTRO-MOTIVE FORCE}}{\text{RESISTANCE}} \text{; or, if}$$

$$(Internal + External)$$

expressed in symbols, C or $I = \dfrac{E}{Ir + Er}.$

Now, in constructing a battery, the *object to be attained*
must be first considered. A battery designed to produce
heat (the cautery battery), for example, is not built upon the
same plan as one designed for ordinary medical purposes.

Again, different cells (such as those devised by Daniell,
Grove, Leclanché, Grenet, Bunsen, Smee, Hill, and others)

possess special *advantages and disadvantages which have to be considered carefully* before a decision is made respecting the one which should be employed.

Finally, the *number of cells*, the **arrangement of the elements**, and the *size of the elements* are problems to be deter-

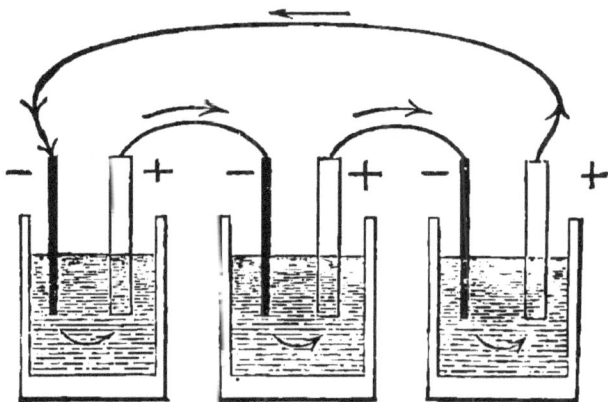

Fig. 5.—A Compound Chain (after Erb).—Three sets of elements are here connected "behind one another," or "in series." The direction of the current is shown by the arrows. The circuit of closure is effected by a wire, as in Fig. 4, *B*.

mined with special reference to the purpose for which the battery is designed. These points will be touched upon hereafter.

It is important that a few facts be stated in the beginning respecting the more common methods of connecting and grouping galvanic cells. Subsequently, the different forms of cells employed by well-known manufacturers of electrical apparatus may be tersely described with advantage. Finally, the various attachments to an electric battery designed for medical purposes should be mentioned, and the uses of each briefly outlined.

Let us suppose, for the purpose of illustration, that we have decided to use a certain number of cells (one of the numerous forms subsequently mentioned) in preparing a battery for medical use. How shall we connect them so as to best accomplish our purpose?

If we join the carbon and zinc elements together (using that form of cell for example), and continue to do so through-

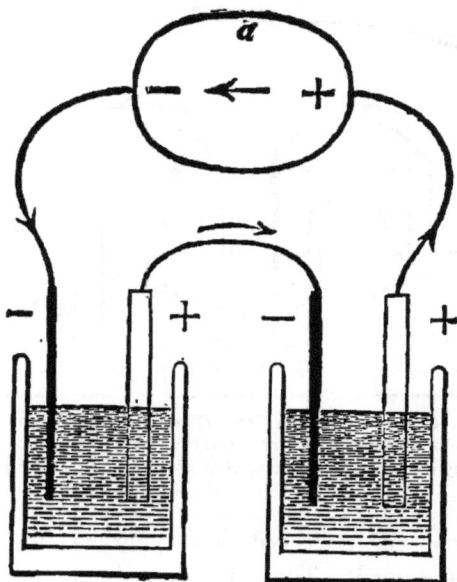

Fig. 6.—A Schematic Representation of the Introduction of a Human Body (*a*) into the Circuit of Closure of a Galvanic Chain (after Erb).— + = the anode ; — = the cathode.

out the entire series of cells (Fig. 7), we have formed what is known as a "*compound circuit*," or an arrangement "in series." If we join all the negative or carbon elements together, and then the positive or zinc elements in a similar way, we have what is known as a "*simple circuit*" (Fig. 8). Finally, we may *divide the cells into groups ;* then join those of each group in simple circuit; and afterward unite these groups as if they were single cells.

Now, what will the effect of each of these methods of combination have on the intensity of the current? Ohm's law comes into play in deciding such a problem.

We must first ascertain the internal resistance of the form of cell which we have selected for our battery.* We must know also the external resistance which we shall have to overcome in our proposed use of it. Finally, we must ascertain the electro-motive force of the elements of each cell.

Suppose, for example, that $E = 1$, $Ir = 20$, $Er = 10$. The current of each cell would then be expressed as follows: $C = \dfrac{1}{20 + 10} = \dfrac{1}{30} = .033 +$. Now, if twenty cells of this kind be joined in "simple circuit," the elements have each been practically increased twenty times, and the internal resistance has therefore been decreased twenty times. The

---

* To compute the internal resistance of a cell or battery requires apparatus not generally owned by medical practitioners, *i. e.*, a coil rheostat, which may be confidently regarded as accurate, and a carefully calibrated galvanometer, by a standard maker. The rule given by De Watteville, and copied from him, apparently, by Amidon, would be simple if it were true. I have tested it again and again, and have personally discarded it as unworthy of credence. I have also had a professional electrician test it. He arrived at the same unsatisfactory results. The rule of De Watteville, to which I refer, is as follows: First note the needle-deflection of the cell or battery to be tested under a given resistance; then introduce sufficient additional resistance to reduce the recorded needle-deflection exactly one half. The added resistance will equal the internal resistance of the cell or battery tested. The internal resistance of any cell can be computed with accuracy; but by a more complicated method, described in most of the standard works upon electricity. Most manufacturers can give the requisite information respecting the internal resistance of any cell used by them, and that resistance, multiplied by the number of cells employed, will equal the total resistance of a battery (the cells being united "in series," as shown in Fig. 7).

external resistance remains the same. We therefore have

$$C = \frac{1}{\frac{20}{20}+10} = \frac{1}{11} = .0909+.$$

If these cells be now arranged in "compound circuit,"

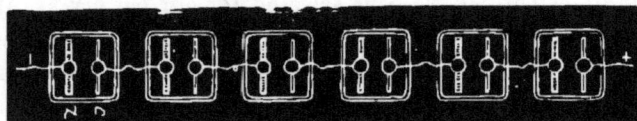

FIG. 7.- SIX CELLS CONNECTED FOR INTENSITY. (After De Watteville.) z, zinc elements; c, carbon or platinum elements. This arrangement is known as "in series" or "compound circuit." It increases the "electro-motive force" of the battery.

the electro-motive force and the internal resistance will be increased twenty times. We should thus have the following formula: $C = \frac{1 \times 20}{20 \times 20 + 10} = \frac{2}{41} = .048+.$

Finally, if the cells were arranged in four groups of five each in simple circuit, we should have practically four cells with elements five times as large; hence the internal resist-

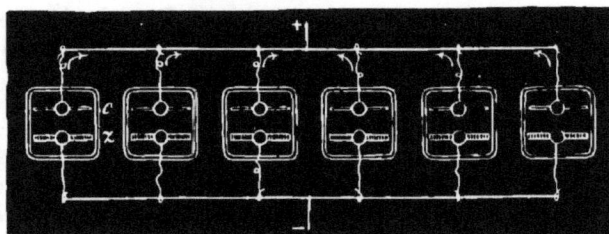

FIG. 8.- SIX CELLS CONNECTED FOR QUANTITY, i. e., "in surface," or in "simple circuit." (After De Watteville.) z, zinc elements; c, carbon or platinum elements. This arrangement does not affect the "electro-motive force" of the battery.

ance would be only one fifth that of a single cell and the electro-motive force four times as great. We should then have the following formula: $C = \frac{1 \times 4}{\frac{20}{5}+10} = \frac{4}{14} = \frac{2}{7} = .285+.$

Remember that the *electro-motive force means the differ-ence in potential of the cell-elements. It is therefore un-changed by their size.* A cup of water elevated one hun-dred feet will produce as much pressure through a pipe connected with it (provided that the cup be kept constantly filled) as would a lake ten miles in circumference, at the same elevation and similarly connected with the pipe. So it is with electro-motive force. The size of the elements will alter the *quantity* of electricity generated ; but the elec-tro-motive force of a cell or battery will remain undisturbed by increasing or diminishing the size of the elements. I frequently hear this remark made : " The cells are too small for medical purposes, are they not ?" To this question I would reply that intensity of current and moderate quantity are to be aimed at in constructing a medical battery.

The few illustrations which have been given show that the current-strength has been modified in each in-stance by the changes made in the arrangement of the cell-elements.

If, however, we took a higher external resistance (as would be required in a medical battery—say about 2,500 ohms), we should find that the simple circuit arrangement made but little difference in the power of the current, while the compound circuit materially increased the current-strength. It is important to remember, therefore, that the *external resistance is an important factor in modifying the strength of the current,* and that all combinations of cells are not equally efficient.

The most useful battery for medical purposes is one which is planned with a view of making the internal and external resistances as nearly equal as possible.

When we wish to construct a battery for *ordinary gal-vanic treatment,* it is best to overcome the high resistances encountered by using a *large number of small cells, with a high electro-motive force, coupled in compound circuit*—i. e., "in series." The aggregate internal resistances of the cells never will exceed the external resistance furnished by the living tissues.

In devising a battery for *electrolysis*, the arrangement should be such as will secure simple intensity. The resistance to be overcome by the current in passing through small portions of the body seldom exceeds 100 to 500 ohms. A *small number of cells of medium size* (16 to 24 of Grenet's cells), coupled in compound circuit, will give us the desired ends and accomplish the best results.

A *cautery battery* requires *very large plates, placed closely together.* In the "Piffard battery" the zinc plates are perforated, and the elements are so arranged as to be mechanically shaken in the fluid while the battery is in action. I regard this as the best instrument of its kind. Its action is continuous, and the heat generated may be maintained at any desired temperature by one familiar with its management.

THE VARIOUS FORMS OF CELLS.—Human ingenuity has been strained to its utmost for nearly a century to devise an absolutely perfect galvanic cell. Space will only allow here of a brief statement of the varieties of cells now in common use. The construction of each and the peculiar advantages and disadvantages of each will be also tersely mentioned.

SPECIAL FORMS OF THE GALVANIC CELL.

All forms of galvanic cells may be classed under one of three groups, as follows : (1) one-fluid cells, with no depolarizer ; (2) one-fluid cells, with a solid or liquid depolarizer ; (3) two-fluid cells.

Each of these three varieties has many modifications, which are commonly named after the inventor. A few of each only need be mentioned.

I. ONE-FLUID CELLS, WITH NO DEPOLARIZER.—The elements of this group are all immersed in a fluid to which nothing has been added to prevent polarization (i. e., the formation of bubbles of hydrogen on the negative and of oxygen on the positive element of the cell during its period of activity).

Smee's Cell (1840).—Perhaps the best of this group is that devised by Smee. It consists of two zinc plates with one of platinized silver, suspended between the zincs, immersed in diluted sulphuric acid. The electro-motive force is about ·47 volt.

Walker's Cell (1859). — Platinized carbon is used in place of platinized silver. It is cheaper than Smee's cell. E. M. F. = ·66 volt.

FIG. 9.—SMEE'S CELL. This is a favorite with some manufacturers for a portable faradaic machine. In the author's opinion, it is far less satisfactory than Fuller's cell if the battery is a permanent one, or a Grenet cell if the battery is designed for transportation. It is active at first, but weakens rapidly from polarization.

II. ONE-FLUID CELLS, WITH SOLID DEPOLARIZERS.—The best of this group is the cell devised by Leclanché.

Leclanché's Cell (1868).—The carbon element is packed in a porous cup, with the needle form of the black oxide of manganese surrounding it. This cup is then placed in a

2

glass vessel, containing a rod of zinc and a solution of sal ammoniac. The cup is carefully sealed to prevent evaporation and escape of its contents. E. M. F. $= 1\cdot48$ volt, when the battery is not polarized.

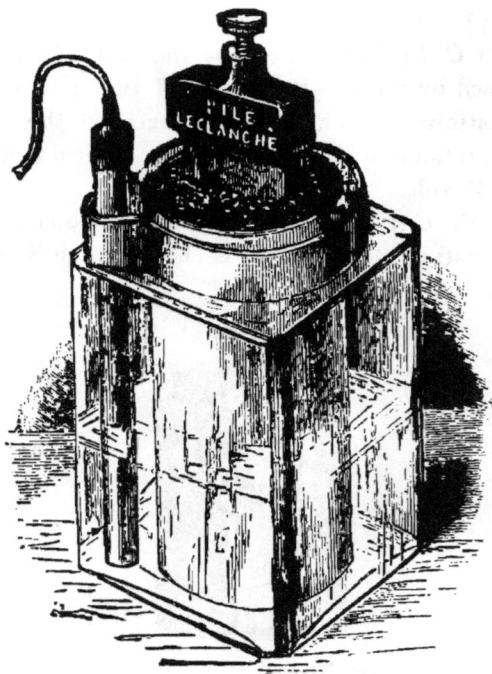

FIG. 10.—LECLANCHÉ'S CELL. The zinc rod (the one with its rheophore attached) is shown as immersed in a solution of ammonic chloride. The carbon element is seen to project slightly above the porous cup, in which, when the cell is properly prepared for action, it is packed with peroxide of manganese and afterward covered with pitch.

Marié-Davy Cell.—Amalgamated zinc, acidulated water, carbon, and a paste of sulphate of mercury. E. M. F. $= 1\cdot52$ volt.

Agglomerate Leclanché Cell.—The carbon is surrounded by plates of a special composition, which are bound around it by India-rubber bands. The internal resistance can be intensified by adding plates as desired. E. M. F. = 1·48 volt. The internal resistance with three plates = 1·8 ohm; with two = 1·4 ohm; with one = ·9 ohm.

III. One-fluid Cells, with Liquid Depolarizers.— Of this group the Grenet cell is the most used for medical purposes.

Grenet's Cell.—The elements are two plates of carbon and one zinc plate (amalgamated). The zinc element can be lowered into the fluid or raised at will. It lies between the carbons. The depolarizer is bichromate of potassium. The active constituent of the fluid is dilute sulphuric acid.

Fig. 11.—Grenet's Cell. The flasks come of all patterns, according to the taste of the various makers. In the form here depicted the zinc is lowered into the fluid by a jointed handle. This is the cell most used in portable electrical apparatus. It is cheap, efficient, and easily repaired. Removing the elements and replacing them overcomes "polarization" in case the current grows weak from that cause. Some makers place the zinc in a drip-cup when the cell is not in use.

These two ingredients form what is known as the " red-acid fluid." These cells are of different sizes.

Trouvé's Cell (1875).—Similar to Grenet's, but of large

size. E. M. F. = 2 volts. The internal resistance varies from ·0016, when first set in action, to ·07 after the "spurt." The plates are raised and lowered by a windlass. The extent of immersion can thus be graduated. This form of element is known as a "plunge-battery."

Fuller's Cell.—A porous cup containing zinc, mercury, and water is placed in a large glass jar containing red-acid fluid, in which a large carbon plate is immersed. The mer-

Fig. 12.—Fuller's Cell. This is the best cell (in the opinion of the author) to use in connection with a permanent faradaic machine. It is not well adapted for transportation.

cury keeps the zinc amalgamated. The elements are not removed when the cell is not in action. This form of cell is perhaps the best one yet devised to run the faradaic part of a cabinet battery.

IV. Two-fluid Cells. — In this group, the Daniell, Grove, and Bunsen cells are the most used. The two latter are not well adapted for medical purposes. The fumes which arise from some of them are unpleasant. Dynamos are now generally substituted for them in the mechanical arts.

Daniell's Cell (1836).—The so-called "sulphate of copper" cells (of various types) are modifications of that devised by Daniell. The elements are zinc and copper, separated in the original form by a porcelain or baked-clay diaphragm. The zinc is immersed in dilute sulphuric acid, and the copper in a saturated solution of sulphate of copper. E. M. F. = 1·079 volt. The solution for the zinc element may also be pure water, salt and water, or a solution of sulphate of zinc.

Siemens and Halske's Cell.—This is a favorite cell for medical batteries in Europe. It is a modification of the

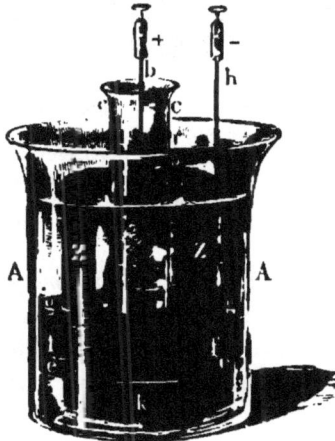

FIG. 13.—SIEMENS AND HALSKE'S CELL. This cell is very efficient, but it is expensive to repair when the battery becomes exhausted. It is highly recommended by some European authorities for use in a cabinet or permanent office battery. In this country the Leclanché cell is more favorably regarded.

Daniell's cell and is expensive. A copper rosette is placed in a saturated solution of sulphate of copper at the bottom of the jar; this is covered with a porous diaphragm packed with *papier-maché*, on which the zinc rests surrounded by

water. Water is added to the battery from time to time,
and also crystals of the sulphate of copper. This form of

Fig. 14.—Hill's Gravity Cell. This cell is employed very extensively in telegraphy, and is recommended by some authors for permanent medical batteries. When the jars are well paraffined at the top the cells do not "salt"
badly. They require but little care when properly set up. Personally, I
prefer a modification of this cell (in which the zinc is placed within a suspended porous cup) to the one shown in the cut. It requires less care, and
is not affected by agitation. It also has a higher internal resistance.

cell is very constant; but it is *extremely difficult to repair*
when out of order. As a permanent battery, such cells may
last a long time with proper care; but they often do not, as
the cells may become impaired from a multitude of causes
(poor construction, improper use, etc.).

Hill's Gravity Cell.—This is another modification of
the Daniell cell. It is used in medicine by many neurologists. A copper plate rests on the bottom of the glass jar,
covered with a saturated solution of sulphate of copper.
The zinc element is a disk perforated by a large central
opening, through which crystals of sulphate of copper may
be dropped when the battery is inactive. A solution of
sulphate of zinc floats, without an intervening diaphragm,
on top of the copper solution, and immerses the zinc disk.

This battery must not be agitated, as the two solutions would then become mixed. E. M. F. = 1·068.

Fig. 15.—GROVE'S CELL. This cell (shown here in the form of a battery) is not used in medical practice, chiefly on account of the fumes which arise from it. If used in the mechanical arts, it is a very expensive cell to employ. Dynamo-machines have now taken the place of Grove batteries to a very great extent.

Grove's Cell (1839). — This consists of amalgamated zinc immersed in dilute sulphuric acid within a porous pot.

Fig. 16.—BUNSEN'S CELL. Parts are represented as bitten away to show its arrangement: the cylinder of zinc (z); a porous cup (r); and the carbon (c) within it. This cell is not employed in medical batteries, for reasons similar to those given in connection with Fig. 15.

Outside of this pot platinum is immersed in nitric acid placed in a glass jar. E. M. F. = 1·96 volt. The platinum is bent into an S-shape to increase its surface. Many modifications of this cell have been made for use in mechanical arts. The fumes arising from it are very objectionable.

Bunsen's Cell (1840).—This is a modification of the Grove cell. The platinum is replaced by artificial carbon in the form of a hollow cylinder, and a cylinder of zinc is bathed in dilute sulphuric acid. E. M. F. = 1·9 volt.

ATTACHMENTS TO A COMPLETE BATTERY.

Although it is not necessary for a general medical practitioner to have all of the attachments to a battery such as are employed by neurological specialists, still it is important that they be mentioned here, and their uses interpreted to you. The most important attachments to a cabinet, or fixed battery for office use only, are a galvanometer, a rheostat, a thermo-electric differential calorimeter, a polarity changer or " commutator," and a variety of rheophores and electrodes. Portable batteries do not require the first three of the attachments mentioned, but they should possess the others.

THE GALVANOMETER.—When a galvanic current circulates in a coil of wire about a magnetic needle, it causes deviations of that needle, which are modified by both the strength and direction of the current deflected into the coil. This fact has led to the construction of an instrument, called "the galvanometer," for the purpose of measuring the *strength* and *direction* of a current deflected into a coil beneath such a needle. When this instrument (properly made and calibrated) is connected with a battery, the strength of any number of cells in milliampères can be determined.* It is vitally important that the dial of a horizontal galvanometer should *not be divided into degrees of equal distances.* Such

---

* Every galvanometer should measure at least quarters of the first milliampère to be considered worthy of indorsement.

a galvanometer is absolutely worthless. The graduation of
the dial should be *by tangents*, as shown on Fig. 18.

The deflection of the needle grows less and less for every
milliampère of current ; hence a dial, to be accurate, should

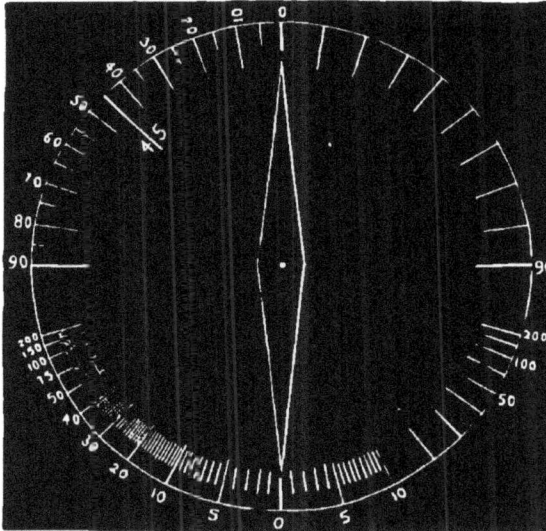

FIG. 17.- A GALVANOMETER DIAL (after De Watteville). The lower half of the
circle is graduated to milliampères ; the upper half to degrees of equal distance. One serious criticism can be made of this dial, viz. : that it does
not indicate fractions of the first milliampère of current. To my mind, a
galvanometer-needle deflection for the first milliampère of current should
be sufficient to show at least a quarter or an eighth of a milliampère. This
fault is common to all vertical milliampère-meters with which I am ac-
quainted ; even Hirschmann's instrument does not entirely overcome it. I
am at work at present upon a new form of milliampère-meter, which I hope
will remedy this serious objection and at the same time allow the needle
deflections to be read easily when the eye is on the same level as the needle.
I do not believe that it is necessary to graduate any galvanometer above 40
milliampères ; no patient will endure a current of over 20 milliampères
through a high resistance, and very few will bear one of 12 milliampères.

be carefully graduated so as to correspond with the needle-
deflections for different current-strengths. The first divisions

on either side of the zero point on such a dial will be coarse, but they should gradually grow finer and finer till the maximum point is reached. *Such a dial will not be gradu-ated around its entire circumference,* as the maximum point will be reached before the 90° of the circle on either side are required. My own galvanometer is graded into *equal degrees* on one half of the dial, and on the other half it is calibrated to milliampères.

Within the past few years the efforts of Erb, Eulenburg, and Bernhard in Germany, Gaiffe in France, and De Watte-

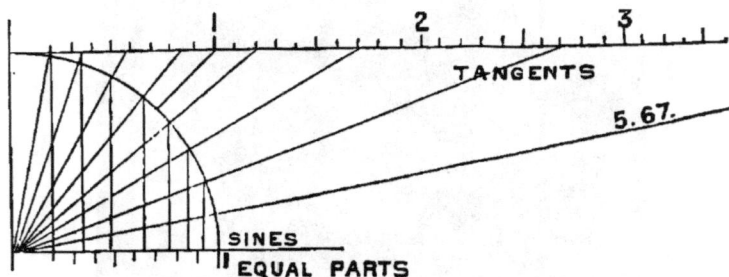

FIG. 18.—A DIAGRAM DESIGNED TO ILLUSTRATE THE METHOD OF TANGENT CALIBRATION. The distances marked upon the straight line are uniform. When they are joined by imaginary lines with the center of the circular dial, these lines intersect its circumference at points which steadily tend to approach each other ; hence the first milliampère will produce a needle deflection which may exceed that produced by ten or more milliampères in some other part of the dial. The more sensitive the needle, the greater will be the distances marked upon the straight line, and the dial also, on either side of the zero point ; hence a very sensitive needle, balanced so as to avoid unnecessary friction, will record eighths and quarters of one milli-ampère of current, if the coil be long enough.

ville in England, have awakened the profession to the neces-sity of accurately measuring the current-strength employed upon a patient by means of a reliable galvanometer. To their views I lend my most hearty support. As well can I conceive of a boiler without a steam-gauge, or of a drug-store without a scale as a galvanic battery without a gal-

vanometer, provided its possessor aims at scientific precision in his treatment of patients by galvanism. Much
of the neurological literature we now possess is materially
lessened in value by the fact that the observations recorded
lack scientific precision. If we expect to arrive at positive
conclusions regarding methods of employing electricity for
therapeutic or diagnostic purposes, we must have a more
accurate and scientific basis for recording the strength
of the current employed than the simple statement of an
observer "that a certain number of cells were used" in each
particular instance. Cells vary in their capacity and electro-
motive force; they change in both respects from day to
day, from use and polarization; the external resistance
afforded by different individuals is not uniform, although
the poles may be similarly placed and all precautions taken
against poor conduction; and many other sources of error
may creep into observations practically made by "guess-
work" only. The scientific world has now quite generally
accepted the "milliampère" as the recognized standard of a
unit of current-strength. A milliampère-meter is therefore
the instrument which each of you should own, and all of
your observations should be recorded from the deflections
of its needle. We shall probably be able soon to state with
some positiveness the number of milliampères which are
required to excite each of the more important nerves of the
human body in health, and the exact limits between which
contractions of certain muscles may thus be excited. Eulenburg and Weiss have already made a step in this direction.

One reason why I prefer the vertical form of galvanometer to the horizontal is the fact that terrestrial magnetism
does not exert an appreciable influence upon it; hence observations made with such an instrument would be alike in all
parts of the globe. The vertical galvanometer is, however,
more expensive, provided it is accurately graduated, than the

horizontal—a fact which, perhaps, would lead me to advise
you to purchase the less expensive instrument.   At my
request, Messrs. Waite & Bartlett have lately graduated

FIG. 19.—A HORIZONTAL MILLIAMPÈRE-METER (after the Thistleton pattern).
The screw-feet allow of adjustment so as to insure a perfect leveling of the
instrument.   It is then revolved so that the needle (which will point north)
rests at the zero point of the dial.   Reversal of the current diverts the
needle to the opposite side.   One of the rheophores shown goes to a binding-post of the battery, and the other to one of the electrodes employed
upon the patient.   This instrument is very delicate, but the eye has to look
down upon the dial in order to observe the deflection of the needle.   If the
instrument is placed lower than the eye, this objection is not serious.   With
it it is easy to detect small fractions of a milliampère of current, and it is
much less expensive than a good vertical galvanometer and more accurate
than most of those offered to the profession.

some horizontal milliampère-meters with great accuracy by
a Thistleton's instrument.   I can recommend them as reliable and comparatively inexpensive.   Before you purchase this important attachment to a battery, test it, if
possible, by one whose accuracy can be relied upon.   I
would far rather have none than a poor one.   At present I am testing a vertical milliampère-meter, made after
a design of Dr. Rudisch.*   Hirschmann's vertical galva-

* This is shown in a cut of my own cabinet battery, on a succeeding
page, and in Fig. 20.

nometer with astatic needles, called the "absolute galvanometer," has been highly recommended of late. I have
not yet tested it.

Fig. 20.—A VERTICAL MILLIAMPÈRE-METER (after an original design by Dr.
Rudisch). This instrument I have tested of late upon my own battery. In
some respects it is an excellent one ; it will not, however, record fractions
of a milliampère unless the dial is magnified by a glass. This objection
holds good, although to a less degree, of the more elaborate instrument devised by Hirschmann, which is provided with "shunts." An absolutely
perfect galvanometer has yet to be devised. Terrestrial magnetism can
easily be overcome, but there are some mechanical difficulties that are more
serious. A vertical needle (if very long) is affected by gravity when it is
deflected by a strong current toward the extreme limits of the scale. A
horizontal needle obviates this difficulty, but is more or less affected by
terrestrial magnetism.

THE RHEOSTAT.—This is an appliance to regulate the
external resistance of a battery under varying circumstances.
Several devices are made for this purpose, but the fluid
rheostat (containing water, solutions of salt, solutions of
zinc sulphate, etc.) is all that is required for purely medical
uses. It is cheap, easily managed, and sufficiently accurate
for the purpose. It is liable to polarize * when used too

long, and does not act so well with strong currents as with
weak ones. It consists of a glass tube filled with water or
some prepared solution (preferably a forty per cent. solution of sulphate of zinc), through which a brass rod or an
amalgamated zinc electrode is made to slide up and down,
thus separating its lower end from a button at the bottom
of the tube. When the current is sent through this rod, it

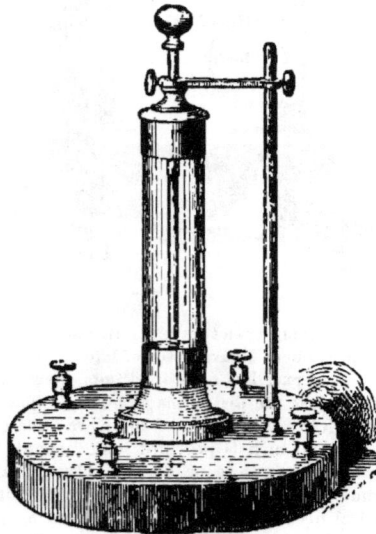

FIG. 21.—A FLUID RHEOSTAT. This instrument is used to throw additional
external resistance into circuit. Coil rheostats are more reliable, because
fluid is decomposed and causes polarization of the metal points when
strong currents are used. Some of the graphite rheostats are preferable to
those containing fluid, and are cheaply made.

is forced to pass through the depth of fluid that lies below
it in order to reach the button at the bottom of the tube.
By moving this rod, the amount of fluid which is thus in-

* We speak of an element as "*polarized*" when bubbles of hydrogen or oxygen accumulate upon it and thus diminish its efficiency.

terposed in the circuit of the battery can be graduated to any desired point. In this way a greater or less resistance can be made at will. A fluid rheostat is absolutely useless for measuring the strength of a current, but is an excellent appliance for modifying it.

THE THERMO-ELECTRIC DIFFERENTIAL CALORIMETER.— This apparatus is used in medicine to measure differences in temperature in homologous parts of the body, or of any two selected points. It is much more delicate than any form of surface thermometer, and is often very valuable as an aid in making a scientific diagnosis. Like many of the instruments of precision, it requires experience to use it and it is somewhat expensive. The study of surface thermometry has not assumed the importance which, in the opinion of the author, it is destined yet to take. Waite & Bartlett, of New York, have constructed for me one of the most perfect instruments of this kind that I have ever seen.

The study of cerebral thermometry has already led to the discovery that the left hemisphere is normally hotter than the right (Hammond); that willed muscular action raises the temperature of the scalp over the motor centers called into action (Amidon, Gray, and others); that mental activity, emotional excitement, and merely arousing the attention cause a rise in temperature (Lombard); and that tumors of the brain or its envelopes are indicated by a localized rise in temperature over the site of the neoplasm. I have lately made some novel and interesting experiments in this field which I propose shortly to publish. In detecting inflammatory conditions of the abdominal viscera, this instrument has lately been employed with satisfactory results. I have lately made an improvement * upon this instrument

* The improvement to which I refer consists in the addition of a polarity changer of my own construction which enables me to reverse the deflection of the needle without removing the thermo-piles from the

FIG. 22.—THERMO-ELECTRIC DIFFERENTIAL CALORIMETER.—Connect the two thermostats as shown in figure, viz. : connect by means of one of the metal tipped cords one binding-post of each of the thermo-piles to the two binding-posts on base of the galvanometer. Then connect the two remaining ports, one on each of the thermo-piles with each other. After so doing, place the thumb on the face of one of the thermo-piles and observe the direction of the deflection of the galvanometer needle, then place thumb on face of the other thermo-pile, leaving the first uncovered, and, if the deflection is in the opposite direction to that first obtained, the instruments are properly connected. If, however, the second deflection is in same direction as obtained by pressing thumb on first thermo-pile, disconnect the two cords from either thermo-pile and interchange them, viz. : take cord from right-hand post and place in left, and cord from left post and place in right-hand post ; the deflections will then be as first alluded to, one pile turning needle in one direction and the other in the opposite direction.

which enables the physician to detect differences in the electro-motive force of the thermo-electric piles employed.

surface of the patient. If the needle, for example, shows a deflection of 15° on the right side and 12° on the left side, the difference of 3°

This corrects all errors in observations made with this instrument.

THE CURRENT-SELECTOR.—This device is now added to all of the modern galvanic batteries, whether designed for office use or for transportation. By it, the *number of cells desired can be thrown into circuit.* If a circular dial studded with buttons (which represent the cell-connections) is used, the bar which revolves and impinges upon the buttons acts as a connection between the button on which it rests and the metallic pivot on which it revolves.

It is important that this bar be broad enough to touch each button before it leaves the one behind it, otherwise the current is apt to be broken when the intensity of the current is changed. I have known of serious results from such an accident when a large number of elements were being used upon the head of a patient. At my suggestion, a modification of the dial current-selector has been made by Waite & Bartlett, of this city (Fig. 2?).

Another form of current-selector is that employed by the same firm for some of their instruments. I greatly prefer it to any other kind for a portable battery, as it enables the operator to select the desired number of cells from any part of the battery—thus insuring an equal amount of wear upon all of its parts. It consists of pins projecting from the dial-plate, each of which represents one cell. These may be brought into circuit by means of two metal caps which are placed upon any of the pins desired; the num-

shows double the imperfection which exists in the thermo-pile thus tested, and the proper registration should therefore be $13\frac{1}{2}°$. Such imperfections in thermo-electric piles are practically unavoidable and can be detected in no other way. As far as I know, this defect has never before been remedied. This has heretofore been the only serious drawback to the differential calorimeter, and the addition of this improvement renders the instrument far more valuable for accurate scientific purposes,

ber of pins between the caps will immediately tell the operator how many cells are being used. An objection to its

FIG. 23.—A NEW FORM OF CURRENT-SELECTOR. This allows of a selection of any desired number of cells from any part of the battery, thus insuring equal wear and many other advantages.

use is that it dispenses with a polarity changer—in case the cells employed are capable of being thus selected.

Fig. 24 shows a portable galvanic battery of this make with the current-selector attachment described. In this form one stopper only is employed, however, and the cells in use are indicated by the numerals placed at the base of each pin. With this arrangement a polarity changer is admissible, and is generally affixed to all so constructed. This is particularly to be desired.

One claim that is made in favor of this device is that

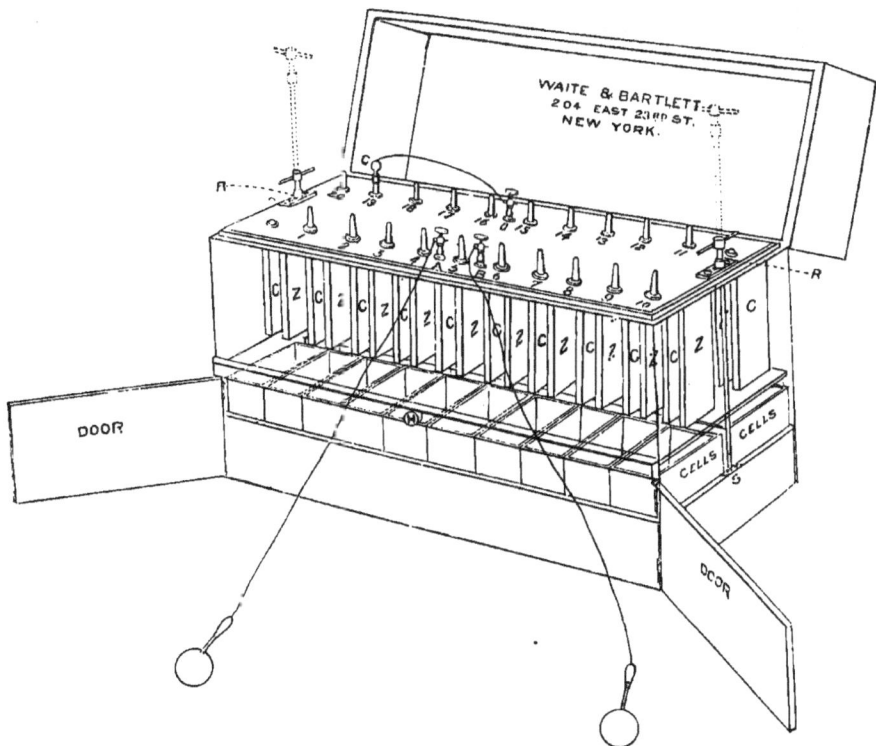

Fig. 21.—A Skeleton Drawing of the Pin Variety of Portable Galvanic Battery. *R*, handles by which the tray of cells is raised and lowered. *Z*, zinc elements. *C*, carbon elements. *A, B*, binding-posts. *D*, attachment of the stopper which fits over the pins. *H*, a rubber-covered diaphragm which separates the cells from the elements when the battery is not in use ; this is removed when the cells are lifted so as to immerse the elements.

This page has been intentionally left blank

this form of key-board prevents oxidation, as the wires which are usually employed to join the cell-elements with the current-selector in other apparatuses are entirely dispensed with. This advantage is an important one to those who live away from the large cities and are not sufficiently familiar with electrical apparatus to make their own repairs.

In batteries which are formed of a large number of cells, it is best to have two dial current-selectors, so that a gradual increase or decrease of the current can be made without breaking the circuit. In one dial, each button should represent one cell, while in the other each should represent from two to five cells. It is easier to make any desired combination of cells with rapidity by means of such an arrangement than if the dials were alike.

THE POLARITY CHANGER OR COMMUTATOR.—Most galvanic batteries have a switch upon the key-board that is intended for the purpose of changing the poles at will without disturbing the rheophores or electrodes. The details of the many mechanical contrivances employed for that purpose need not be given here.

This attachment is almost indispensable to a battery designed for office or experimental work, since the reactions of the poles can thus be more readily studied. It is desirable, moreover, to have it attached to a portable galvanic battery. It should be so arranged as to permit of *opening and closure of the circuit*, as well as the *reversal of the current.*

THE RHEOPHORES.—These are flexible wires which are necessary to conduct the electric current from the battery when in action to the patient. Insulated copper wire forms the best rheophore, as it is an admirable conductor. Tinsel threads insulated with cotton wrapping are more generally used, because they do not kink and are more flexible (although they are not such good conductors). They are connected at one end to the "*binding-posts*" of the battery, and at the other to the electrode. They should vary between four and six feet in length. I have some that are ten feet in length, which I employ when I examine the naked body of a patient lying upon a sofa or bed.

THE ELECTRODE.— In order to apply a current of electricity generated in a battery to the human body, various forms of electrodes are employed as termini to the rheophores. It is best to have a pair of handles to which different forms of tips may be screwed, according to the requirements of each case. The tips may be made of plain metal, or of carbon or metal covered with sponge, chamois-skin, or canton-flannel. The canton-flannel covering is the cheapest and cleanest, and may be renewed at pleasure. Each patient can thus have a clean covering for the electrode at every application. Flat electrodes of large size are useful, especially when a neutral point for the current is desired). Small tips (motor-point electrodes) are generally employed to direct the current to some special muscle or group of muscles.

The wire brush is employed chiefly in cases where anæsthesia exists. It is the only electrode that is used dry.

Most manufacturers advertise a case of electrodes designed especially for the application of electricity to different organs. Selections may be made from these as desired. I have personally devised several modifications of electrodes. These are manufactured by Waite & Bartlett.

A practical point may be mentioned here, viz., that the negative electrode (*cathode*) is the most painful to the patient, and produces the greatest chemical action. It is a well-recognized fact that a bullet does the most damage at its point of escape from any dense substance which it has penetrated. In the same way an electric current produces the most profound effects at its point of escape from the body—i. e., the negative pole. It is not uncommon to see a reddening of the skin, and even vesication, produced by a

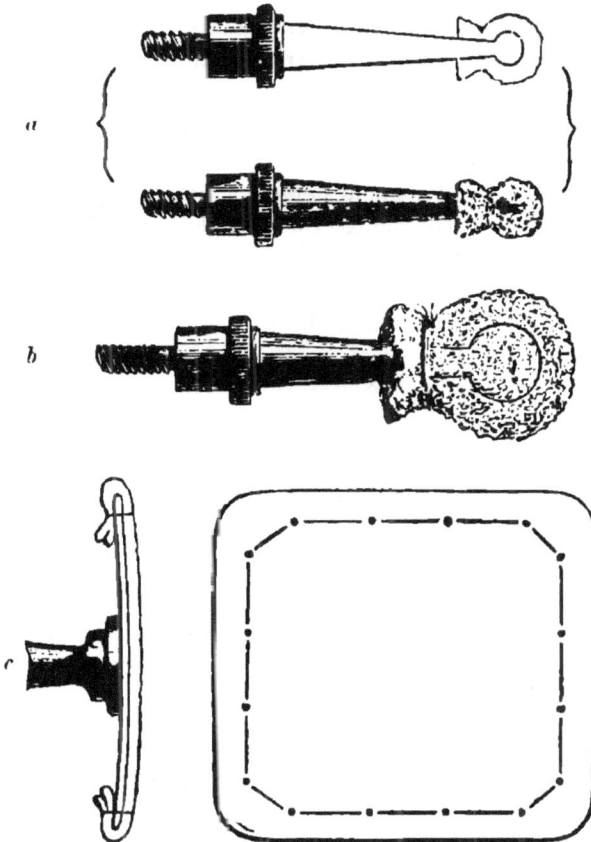

Fig. 25.—Various forms of electrodes (natural size), adapted to a screw handle, not shown in the cut. (After Erb.) *a*, the "fine" electrode or smallest size; *b*, the so-called "small" electrode; *c*, the "medium" electrode. All electrodes are covered with sponge (as in *a* and *b*) or flannel or chamois skin (as in *c*).

strong current at the negative electrode if kept too long in contact with it. The cathode is the "stimulating" or "irri tating" pole (if such an expression is admissible) of a galvanic battery. The anode is, by contrast, the "sedative" pole.

For the treatment of special forms of disease by electricity, different types of electrodes have been designed by various neurologists. A device of my own, which simplifies electro-diagnosis and enables a medical observer to watch and compare the effects of electrical currents of definite strengths upon muscle- and nerve-reactions of opposed limbs simultaneously, will be described farther on. By means of this device the physician may sit at a key-board and excite different sets of muscles separately or simultaneously (without moving from his chair) by touching certain keys, as if

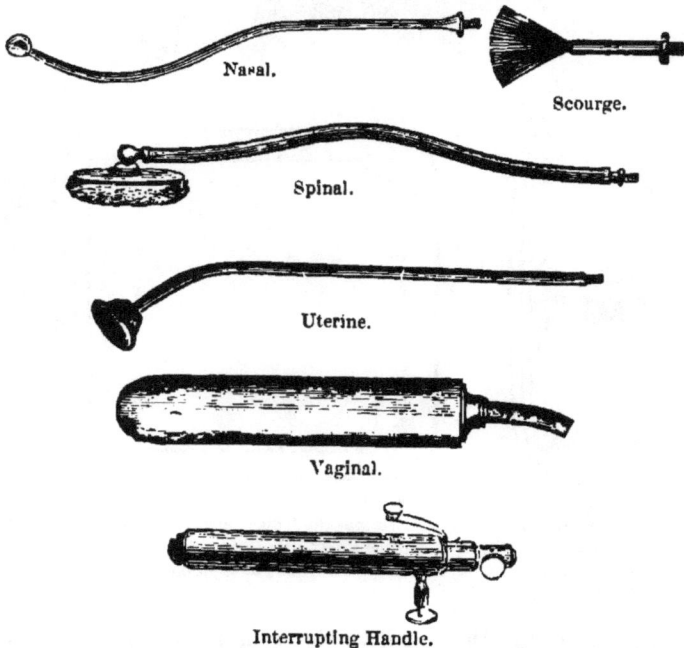

Nasal.

Scourge.

Spinal.

Uterine.

Vaginal.

Interrupting Handle.

FIG. 26.—VARIOUS FORMS OF SPECIAL ELECTRODES.

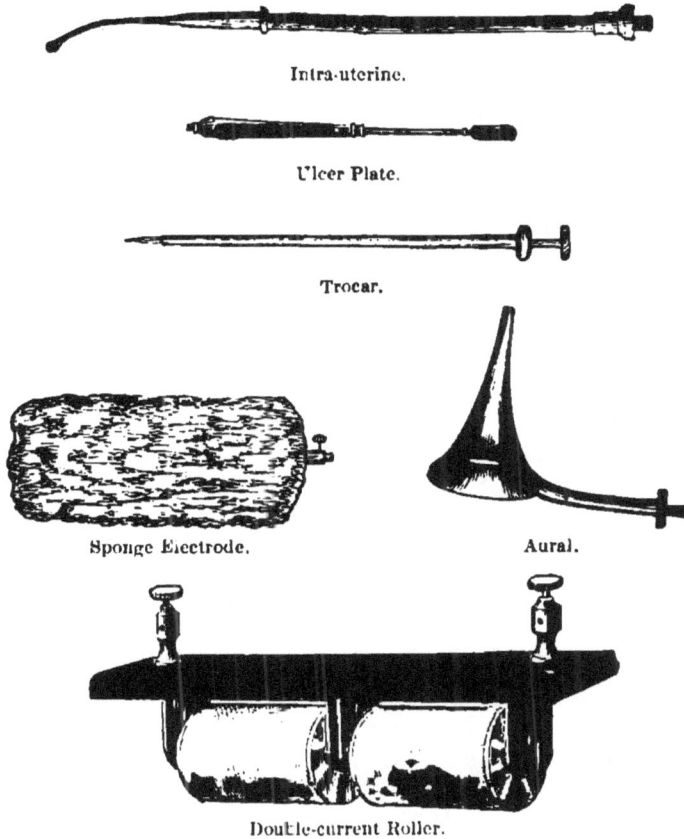

Intra-uterine.

Ulcer Plate.

Trocar.

Sponge Electrode.   Aural.

Double-current Roller.

Fig. 26.—Various Forms of Special Electrodes.

playing the piano or working a type-writer. I have found it, moreover, of great assistance in demonstrating before large audiences the action of nerves and muscles upon a living subject.

Time will not allow of a detailed description of the electrodes shown you, most of them being simple devices, which really require but a limited experience to use them in a proper way.

When you purchase a battery, two sponge-covered electrodes will probably be sent with the instrument. It is advisable, for the following reasons, to remove the sponges: 1. Cases have been reported where disease has been transmitted by sponge-covered electrodes. 2. In case the metal electrodes become oxidized, they can be readily cleaned. 3. When employed upon a patient's body, absorbent cotton, wet in salt water, can be placed upon the clean metal, and a piece of moistened canton-flannel may then be wrapped over both and fastened to the handle with a rubber band, thus insuring *absolute cleanliness* and *perfect conduction of the electric current.*

Patients of delicate sensibilities rebel against the use of the sponges which for months or even years have been employed in case after case requiring electricity. Who of you would patronize a barber-shop where one towel constituted the entire outfit of linen? Is it right to ask of patients what you would yourself condemn? Furthermore, how can electrodes covered with dirt and other deposits under a sponge be perfect conductors of electricity? Absorbent cotton and canton-flannel are far cheaper than sponges, and can be thrown away after being used.

THE CHOICE OF A BATTERY AND ELECTRICAL APPARATUS.

In selecting a battery for purely medical purposes, the chief objects to be attained are (1) *cheapness;* (2) *constancy of the elements,* and their *accessibility for repairs, cleaning,* etc.; (3) *durability of the elements;* (4) a *sufficient number of elements;* (5) *ease of renewal of the elements;* (6) *ease of introduction of any number of elements into the circuit;* and (7) an ability to select such as may be required *from any part of the battery.*

Fio. 27.—The Physician's Handy Cabinet Battery.—The accompanying cut represents a light and compact form of cabinet battery, designed by the author. It is on castors, and can be wheeled about the consultation-room. This admits of the use of the instrument when the patient is in the gynæcological chair or upon the office lounge; or when any form of fixed apparatus, such as the laryngoscope, the inhaler or spray, etc., is being simultaneously employed. In some of my later models an immovable tray placed beneath the battery for electrodes, and a movable shelf is also provided upon which a milliampère-meter, the solution of table salt, and the electrodes in actual use can be set. A glass cover protects the battery from dirt when not in use. E, faradaic coils; K, plunger; G, faradaic binding-posts; F, interrupter; D, drip-cup; R, current-selector of single cells; S, the same of four cells to each button; M, coil to work the interrupter for the galvanic current; L, switch to work or disconnect the interrupter (V); P, galvanic binding-posts; 3-5 and 4-6, connecting rods to allow of the action of M. The commutator lies above P.

3

For the general practitioner it is necessary, as a rule, that a galvanic or faradaic battery *shall be arranged for transportation ;* hence the cups which hold the fluid should have a rubber cover, or some other device which will preclude the possibility of spilling the fluid. Again, some of the batteries manufactured are liable to become rapidly oxidized by the fumes of the battery fluid. This tends to destroy their durability, and to cause difficulty in keeping them in good working order. Finally, it is very desirable that portable batteries should be as light as possible, and not too large to be handled easily.

The *attachments upon the key-board* of every portable galvanic battery should comprise a current-selector and a commutator. There should be at least four rheophores, in order to make allowance for breakage, additional connections, etc. Several electrodes of different sizes and shapes should also be selected—preferably a large, a medium, and a small one—a wire brush, and an interrupting electrode. These can be added to as circumstances demand.

For office purposes a CABINET BATTERY has some decided advantages over a permanent one placed in an adjoining closet or cellar and connected, by means of wires, with a key-board in the consulting-room. A cabinet battery can be easily wheeled about, and is readily repaired. The cabinet should be so arranged as to allow the back and front of the compartment for the cells to be removed, in case the battery needs repairs, or a renewal of the fluid. The connections of the battery with the key-board should also be made as easy of access as possible ; this decreases the expense of alterations or repairs when such become necessary. They should be protected, moreover, as far as possible, against oxidation and dirt.

The cabinet battery which I use in my own office was made, under my special direction, by Waite & Bartlett, of

Fig. 28 — LARGE CABINET BATTERY. From a photograph of one used by the author, and constructed from designs specially furnished by him. The current-selector and rheotome differ in several respects from those commonly used. The faradaic attachment has a Du Bois-Reymond coil. The milliampère-meter shown in the cut is that devised by Dr. Rudisch. I am at present engaged in the construction of one of a novel pattern, which I hope to present to the profession hereafter. The cells are of the Leclanché pattern. The faradaic attachment is operated by a Fuller's cell.

this city, and is as nearly perfect as one could desire. It contains *forty cells of the Leclanché pattern*, which are equivalent to sixty of the gravity cell. The connections and the cells can be exposed and easily reached by removing the front and back of the case. The accompanying cut represents its special features better than a verbal description. Considerable expense in constructing such a battery may be saved in the case, and by dispensing with some of the accessory apparatus shown.

The *gravity cell* makes a very serviceable and durable permanent battery for office work. It has one advantage over some other cells—viz., that it has great constancy of action and that its activity can be renewed by the addition of crystals of sulphate of copper to the fluid when necessary without disturbing the cells. For this reason the sulphate-of-copper cell, in some one of its various forms, is employed exclusively in telegraphic lines. It can not be transported, however, about the room to suit the convenience of the patient or the physician during his examination so well as some other cells adapted for a cabinet battery. It is also difficult in many cases to repair the wire connections of a fixed battery (running, as they often do, through partitions and plastered walls to reach the keyboard) when they become inefficient from any cause.

A PERMANENT BATTERY is somewhat cheaper to construct and takes up less room in the office than a cabinet, because no case is required; but, in my opinion, these two advantages are not sufficient to render it preferable to the other for office or experimental work. I have known several of my medical friends to discard it (after a thorough trial) for a cabinet battery. If a permanent battery is deemed preferable by any of you (for reasons of your own) rather than a cabinet battery, be sure and place your cells on shelves in your office or waiting-room, and not in a cellar. The wires will not be so liable to corrode from dampness, and the cells will be constantly under your eye, so that you can see when they require attention.

Respecting the selection of the cheaper forms of batteries for general medical use, it is important that accuracy of workmanship shall not have been sacrificed in order to lessen the cost. The construction of the primary and secondary coils of a faradaic machine and the adaptability of the interrupter to slow and rapid breaks in the circuit should be looked into before a decision is reached. Poor coils and a bad interrupter render a faradaic machine almost worthless. A "*drip-cup*" containing mercury, in which the zinc element is placed when the battery is not in use, is a desirable feature in a faradaic machine.

Do not buy a magneto-electric machine whose motor power is furnished by a crank to be turned by the hand. It is practically useless for medical purposes when compared with a good faradaic instrument.

The Grenet is cell now used by most of the manufacturers of electrical apparatus for a portable galvanic or faradaic battery. It is the best cell for many reasons. A thirty-cell galvanic battery gives all the current that is required by the general practitioner. Personally, I prefer the one made by Waite & Bartlett, of this city, over that of other manufacturers, on account of its modified current-selector (Fig. 23). It does not oxidize as do other forms of batteries (which have a dial current-selector attachment) when in constant use. It is also cheaper than those made by many other firms. Every galvanic battery should have a *commutator* on the key-board. Without this appliance electro-diagnosis becomes difficult.

Respecting the purchase of a STATIC ELECTRIC MACHINE, it may be well to state that a good one is quite expensive, and is only adapted for office use. I am convinced that static electricity has some points of advantage which can justly be urged in its favor as a therapeutical agent, but it can never be extensively employed or take the place of galvanic and faradaic currents. Its use unquestionably creates a profound impression upon the mind and body of the patient. He sits upon an insulated stool, sees the "wheels go round," feels himself getting charged with electricity, and is made painfully conscious of its presence when sparks of an inch or more in length are elicited from his surface and through his clothing. How much of the reported benefits derived only from the use of this instrument are due to the mental impression so made upon the patient is still a problem which I have not solved to my satisfaction.

The best American instrument of this kind is probably made by I. & H. Berge, of New York. It works well in all weathers, and their largest machine will produce a spark eight inches long. The electrodes for static electricity have to be made specially for its use. They must be well insulated by means of glass or hard rubber. Dr. W. J. Morton * has done much to popularize the use of static electricity in this country. A water-motor is required to run a large static machine with uniformity, although it is not absolutely essential to its use in medical practice, as hand-power will answer all practical purposes. A single-plate machine is but a toy as a means of treatment of nervous diseases. *Sufficient quantity*, as well as length of spark, is essential to the satisfactory employment of static electricity. My own machine consists of six revolving twenty-inch plates and three stationary plates. It works well in all weathers, and gives as large a quantity as any patient can bear.

The following deductions express my convictions regard-

* "Medical Record," April, 1881.

ing the therapeutical value of static electricity, derived from an experience with it in quite a large number of cases:

Fig. 29.—IMPROVED HOLTZ STATIC MACHINE (American Pattern). This form of machine is the best now offered to the profession. The author is at present engaged in devising some radical improvements upon the instrument which will (in his opinion) greatly enhance its practical utility and reduce its cost.

1. In certain diseased conditions I regard its effects as more instantaneous and satisfactory than those of galvanism or faradism.

2. A machine so inclosed in glass as to prevent the

action of dampness upon the revolving and stationary plates
will work in all weathers, and is therefore free from the
most serious objection which can be raised against franklinism.   In the summer months the dampness of the air is
liable to cause diffusion of the electricity generated by friction, and thus to render its employment upon a patient difficult and often unsatisfactory.

3. Quantity, as well as intensity, is requisite; hence
large plates, and several of them, are preferable to any of
the single-plate machines.

4. In the treatment of muscular pains, chronic muscular rheumatism, spasmodic affections, the functional nervous diseases, and neuralgia, I have found this form of
electricity of great value.

5. In treating muscular pains and muscular rheumatism,
spasm, and neuralgias, I prefer the spark to insulation.
Patients who may have suffered for years are frequently
cured in a few sittings.

6. I prefer insulation over all other methods when the
tonic action of static electricity is desired.   Cerebral hyperæmia and anæmia, headache, and vertigo are often
rapidly relieved by this method.   I frequently combine
insulation with the "electric wind" (drawn from the head
by means of an umbrella electrode) in these cases, as the
accumulated electricity is thus concentrated toward the
head.

7. Sparks are particularly to service in treating numbness and cutaneous anæsthesia.   I have remarked this effect
especially in the sensory disturbances which frequently
accompany hemiplegia of cerebral origin.

8. Hemiplegia and paraplegia are best treated by means
of the direct spark rather than the indirect spark.   Sometimes quite severe shocks are required before the remedial
effects become apparent.

9. Wooden electrodes are preferable to those composed of metal when employed about the eye or ear, or when the patient is very sensitive to electric currents.

Fig. 30.—Drawing the Indirect Spark from the Body of a Patient. The chain attached to the electrode is connected with a neighboring gas-pipe or water-faucet. This cut should be compared with Fig. 3.

10. By drawing sparks from the region of weak or diseased viscera, I have sometimes noticed very apparent benefits. Pulmonary, gastric, and hepatic disorders are often directly affected by this agent when so applied. I have relieved bronchitis in this way, and have had in some instances equally beneficial results in thus treating nervous dyspepsia, gastralgia, torpidity of the liver, etc. Some authors have reported beneficial effects from this agent in the treatment of phthisis.

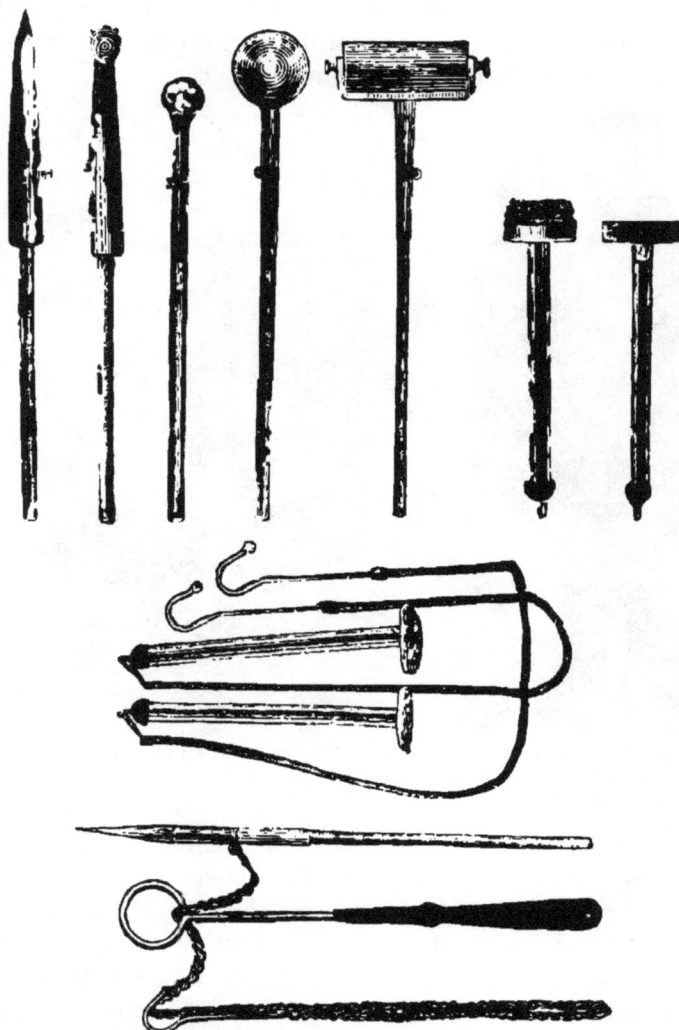

FIG 31.—VARIOUS FORMS OF ELECTRODES EMPLOYED WITH A STATIC MACHINE.
The handles are usually of glass. The author has substituted handles of
hard rubber, which do not break easily and are equally efficient for the purpose of insulation.

11. Muscular contractions can be excited by static electricity with far less pain than by faradism or galvanism. The so-called "static induced current" * is a very efficient way of subjecting individual muscles to the current when their contraction is desired. The painlessness of this method is a point which alone should strongly recommend it.

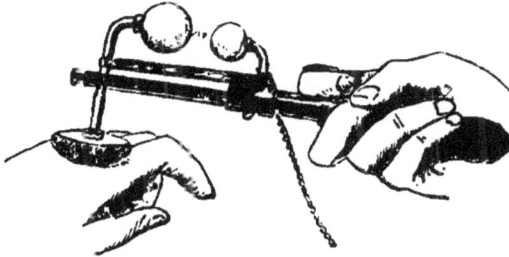

FIG. 32.—MORTON'S SPARK ELECTRODE. The sponge-covered tips may be of any size or shape.

12. Perhaps the most phenomenal results which have been obtained by static electricity are achieved in the treatment of hysterical patients. Hysterical aphonia, hemianæs-thesia, paralysis, and hystero-epilepsy have been cured by a few sittings, by means of insulation and sparks.

THE CARE OF A BATTERY.

THE best battery is liable to get out of order. It is an easy matter, as a rule, to correct the trouble if the construction of the apparatus is thoroughly understood. The following hints may aid the reader in obtaining a satisfactory current with a minimum expense:

1. *Keep your battery clean and bright in all its parts.* Close the case when the battery is not in use, and thus keep out dust, grease, and moisture. Emery-paper is useful to

* First described by Dr. W. J. Morton.

keep the metal connections free from rust. Remember that dirt, grease, or rust will often arrest the action of any battery.

2. When the battery fails to act properly, examine the cells first and see if *the fluid requires renewal.* The " red-acid fluid " is easily made by adding one part of commercial sulphuric acid to ten parts of cold water ; when cooled, one part of finely pulverized bichromate of potassium should be added and mixed well. This is the fluid commonly employed in portable batteries with cells of the Grenet pattern.

3. If the fluid is found to be fresh, and if the zinc and carbon elements are in good order and the zinc well amalgamated, *examine carefully all the screws and other connections attached to the elements* and see if they have become oxidized. Sometimes they become rusted or so covered with accumulated dirt as to render the passage of the current impossible. Occasionally the carbons may be disconnected and baked in an oven to render a Grenet cell more active. Soaking the elements (*in situ*) in hot water which does not reach the connections will generally suffice to cleanse them.

4. If the *cell has become polarized* when in action (by bubbles of hydrogen which accumulate upon the carbon and of oxygen upon the zinc element), lifting the zinc out of the fluid and replacing it immediately will suffice to overcome this trouble if the cells are of the Grenet pattern. These bubbles of hydrogen and oxygen set up a counter-current in the cell which will weaken and may even neutralize the original current.

5. *Examine the interrupter,* the *buttons of the current-selector,* and *the commutator* for rust or dirt, and clean each thoroughly when the trouble appears not to be due to the elements or their immediate connections.

6. If a *drip-cup is furnished* with a faradaic or galvanic battery, be careful to place the zinc element in it when the battery is not in use.

7. In portable galvanic batteries, be sure to *place the rubber-covered diaphragm over the cells* before closing the case and to screw it down tightly. This prevents the fumes rising and oxidizing the connections of the elements when not in use. This does not apply to a red-acid battery when not being transported.

8. Be sure that the *rheophores are perfect* before they are used upon a patient. The wire used in their manufacture is liable to become broken or oxidized by use. This is especially true of the flexible, cotton-covered cords generally furnished with batteries. The electrodes may be tested by employing a galvanometer, if an imperfection is suspected and can not be found.

9. The *wires* that run from the cups to the buttons of the current-selector or the commutator may be seen on the bottom side of the key-board of a battery. They can be examined for imperfections when the other parts of the apparatus appear to be perfect.

10. *Do not short-circuit a battery.* By this we mean, do not allow a battery to run down, or, more technically, "polarize," by the poles being brought into contact without an interposed body (such as animal tissue) for any length of time. For example, galvanic cells which have a low internal resistance (as a Grenet cell) become polarized in a few hours when the poles are connected by a short wire which affords little if any resistance to the current.

11. *Keep your electrodes clean.* It is well to cover them with fresh canton-flannel for every patient. This is an act of precaution which will impress people with your regard for their feelings and for their safety from contact with infectious matter, Sponges are too expensive to be renewed

so often. Absorbent cotton may often be placed between the electrode and its covering with advantage.

Part II

THE PRINCIPLES OF ELECTRO-DIAGNOSIS.*

The various electric tests that are employed as aids in the diagnosis of nervous affections are too complex to be fully described and explained without entering somewhat into the domain of physics and physiology. Erb† has lately written an excellent work upon the subject, and most of the later treatises upon physiology will afford you general information respecting the reactions of healthy muscle to the faradaic and galvanic currents. The few practical hints which are given here are offered with an apology for their incompleteness, although it is hoped that they will assist you in your studies in this field.

Having first moistened the electrodes and connected them with the battery in action, it is customary to hold them both in one hand (close together, but not in contact), and apply them to the ball of the thumb of the opposite hand or the cheek to see if the current is passing properly. If the current to be employed is a *very weak one*, touch the electrodes to the tip of the tongue before it is used upon the patient.

Next, sponge the part of the patient's body to be tested with a *weak solution of table-salt in warm water*, in order to render the skin a good conductor of the electric currents. If the wire-brush is to be used, this step is omitted.

The "*polar method*" is the one commonly used. Apply

* Portions of this lecture have already been published.
† "Handbook of Electro-Therapeutics," New York, 1883.

one electrode of large size, either over the breast-bone of the patient (at about its center) or over the back of the neck. The breast-bone is the preferable point on account of the absence of muscles in the median line.* The other electrode (of small size) is placed over some special nerve-trunk or the muscle to be tested; in case muscle is to be tested, the electrode is placed usually at the point where the motor nerve enters its substance—the so-called " motor-point" of the muscle. In this way the action of the two poles can be readily distinguished.

In my work, "The Applied Anatomy of the Nervous System," I have reproduced von Ziemssen's cuts, illustrating the situation of the motor-points of the various muscles. In case the interrupted or faradaic current is to be employed, the " polar method" need not be strictly adhered to, as it is decidedly more painful than when the electrodes are less widely separated.

Use both the *continuous or galvanic current* and the *interrupted or faradaic current* in testing muscular reactions. The former is of the greatest value in diagnosis.

In *studying the muscular reactions to the different currents* employed, remember (1) that the negative pole is called the cathode (C),† and the positive pole the anode (A); (2) that muscular contractions occur both when the current is altered in strength and when the circuit is closed or opened; (3) that the faradaic current produces an apparently continuous muscular contraction, because its interruptions are so very rapid; (4) that very weak currents do

* This is known as the "indifferent point," when *polar effects* are being studied at the other electrode.

† German authors employ different symbols from those given. These are as follows: C. C. C. = Ka S. Z., C. O. C. = Ka O. Z., A. C. C. = An S. Z., A. O. C. = An O. Z. The symbols Ka = cathode, An = anode, S = closure (*Schliessung*), O = opening (*Oeffnung*), Z = contraction (*Zuckung*).

not produce contractions; (5) that alterations in the strength
of the current cause proportionate variations in the contractions; (6) that the contractions are short, sharp, and sudden in health; (7) that the effects of applying the electrode
over the substance of the muscle and over its motor-point
are identical in health, but not in some diseased conditions;
(8) that the galvanic current will not usually produce muscular contractions while it is constant, but only when its
strength is modified or when the circuit is closed or broken;
(9) that the direction of the current can be changed, without altering the position of the electrodes, by a simple apparatus that changes the cathode into the anode, and *vice versa*
(the *commutator*).

The current passes always from the anode to the cathode.
Hence, when the positive pole is placed on the breast or
neck, and the other on the muscle to be tested, we have a
descending current. An *ascending current* exists if the cathode is on the same distant or neutral point.

An "automatic interrupter" on an "interrupting electrode" is necessary in employing the galvanic current in
testing muscular reactions.

The *descending current,* when closed and again broken,
can thus give us:

1. The cathodal closure contraction:
        C. C. C. or Ka S. Z. of the Germans.
2. The cathodal opening contraction:
        C. O. C. or Ka O. Z. of the Germans.

The *ascending current,* when closed and again broken,
can give us:

1. The anodal closure contraction:
        A. C. C. or An S. Z. of the Germans.
2. The anodal opening contraction:
        A. O. C. or An O. Z. of the Germans.

These four forms of contraction require currents of
different strengths to produce them. They are, therefore,
induced by *gradually increasing the number of cells* employed. The following order is the one commonly observed
in healthy muscle :

    1................... C. C. C. = Ka S. Z.
    2.... .... A. C. C. = An S. Z.
    3................... A. O. C. = An O. Z.
    4................... C. O. C. = Ka O. Z.

It will be observed that the *cathodal contractions* appear
first and last in health, while the *anodal contractions* follow
each other; also, that the *closure contractions* precede the
opening contractions of both the cathode and anode. When
a nerve-trunk is stimulated by electric currents the formula
of the normal muscular contractions is altered. This will
be spoken of hereafter.

Again, as the strength of the current is gradually increased, the contractions which have successively appeared
become intensified proportionately (as is shown below), and
new reactions are added :

First stage (*moderate current*), C. C. C.
Second stage (*stronger current*), C.' C.' C.' and A. C. C.
Third stage (*still stronger current*), C." C." C." and A.'
    C.' C.' and A. O. C.
Fourth stage (*very strong current*), C.''' C.''' C.''' and A."
    C." C." and A.' O.' C.' and C. O. C.
C.''' C.''' C.''' is called "*cathodal tetanus*," because the
    contraction is very violent. Sometimes the anodal con-
    tractions both occur with the same intensity of current,
    thus merging the second and third stages into one.
    Again, A. O. C. may in some cases appear before A.
    C. C.

Disease of the nerve-centers or of the nerves themselves

may cause modifications of the normal formula of muscular contractions. This constitutes the key-note to the value of electric currents in diagnosis. Mechanical devices may be employed to trace the muscular contractions, as the sphygmograph does the pulse. (Fig. 33.)

FIG. 33.—CURVES OF CLOSURE CONTRACTIONS IN DIRECT (UNIPOLAR) STIMULATION OF THE MUS-CLES IN THE DISTRIBUTION OF THE PERONEAL NERVE IN THE LEG (Erb). Ka = C. C. C.; An = A. C. C. 1, Curve of health, thirty-three elements; C. C. C. is greater than A. C. C.; 2, case of chronic anterior poliomyelitis, showing reaction of degeneration with thirty-three elements; 3, same case, with forty elements. In 2 and 3, the excess of A. C. C. over C. C. C. is apparent; 3, in the slow character of the contractions is very marked.

If the destructive process is within the brain or spinal cord, and *situated above the " trophic centers "* of the nerves supplying the paralyzed muscles, the electrical reactions of the paralyzed muscles will be normal in respect to the sequence and character of the muscular contractions. Sometimes, however, a *much stronger current* (galvanic or faradaic) is required to produce them than over the healthy muscles. This fact is due to atrophic changes in the muscles. It may be of decided value in diagnosis.

When disease processes in the brain or spinal cord cause *destruction of the trophic centers* of the nerves that supply the paralyzed muscles, or when the *cerebro-spinal nerves themselves are seriously injured*, we encounter what Erb has described as the "REACTION OF DEGENERATION." This will require some explanation.

1. *Every nerve degenerates when separated from its trophic center ;* hence, the electric excitability of the *nerve*, both to the faradaic and galvanic currents, gradually diminishes and ceases entirely at the end of about two weeks.

2. The *faradaic current ceases to cause muscular contractions* when applied directly over the substance of the muscle. This is explained by the fact that the so-called "nerve-plates" within the substance of the muscle are degenerated, and currents of momentary duration fail to affect the muscular fibers.

3. The muscular contractions produced by the *galvanic current are diminished for about ten days.* Subsequently the excitability of the muscles *to slowly interrupted galvanic currents becomes increased*, so that very weak currents may excite contractions. This may disappear in five or six months.

4. The *polar reactions become altered in their sequence.* The anode contractions appear before those of the cathode, as shown below :

1 . . . . A. C. C. instead of C. C. C. as in health.
2 . . . . C. C. C.   "   A. C. C.   "   "
3 . . . . C. O. C.   "   A. O. C.   "   "
4 . . . . A. O. C.   "   C. O. C.   "   "

5. The *character of the muscular contractions becomes altered.* In health, they are sharp, short, and sudden.

When degeneration occurs, they are slow to appear; they are prolonged and continue even during the passage of the current; and, finally, they assume the character of "tetanic" contractions, irrespective of the strength of the current employed.

Finally, in unilateral paralysis the electrical reactions of the muscles of the paralyzed side should be contrasted with those of homologous muscles of the unaffected side. When both sides are impaired, the standard of comparison should be that of a healthy subject of about the same size, weight and muscular development.

Now let us suppose that we wish to test the reaction of some special nerve—the musculo-spiral, for example. We place the positive pole (An) of a galvanic battery over the breast-bone with a large flat electrode attached, and the negative pole (Ka) over the nerve (where it winds around the humerus below the deltoid muscle) with an "interrupting" small electrode attached to the negative rheophore. We then put into circuit a few cells at a time and press the button of the interrupting electrode at intervals till we get a contraction of muscles. When the current is sufficiently strong to excite the nerve-trunk, contraction of the extensor muscles of the forearm becomes apparent (the *cathodal closure contraction*). Thus we ascertain the number of cells required to produce C. C. C. (Ka S. Z. of the Germans). Now add a few more cells, and reverse the poles by means of the commutator. When the circuit is *broken*, by releasing the button of the interrupting electrode, we get the anodal opening contraction (A. O. C., or An O. Z.), and, with a few more cells, the anodal closure contraction (A. C. C., or An S. Z.). Again reverse the current, and add a few more cells. Now, on pressing the button of the interrupting electrode, we get a very intense cathodal closure contraction (C'''. C'''. C'''., or Ka S. Z'''.), and, on releasing it, the cathodal opening contraction (C. O. C., or Ka O. Z.) is developed, thus completing the chain of polar nerve-reactions.

You should bear in mind that the *polar nerve-reactions differ in their normal sequence from those of the muscles* when the electrode is placed over the "motor point" of the muscle tested.

NORMAL NERVE-REACTION.

C. C. C. > A. O. C. > A. C. C. > C. O. C.

NORMAL MUSCLE-REACTION.

C. C. C. > A. C. C. > A. O. C. > C. O. C.

The *final contraction* (C. O. C.) of each of these series is seldom seen, because the current required to produce it is too painful to be endured. *Fewer cells are required to cause muscular formulæ than those of a nerve-trunk.*

In recording the results of an electrical examination of nerve-trunks and muscles it is best to arrange the record-page so that the two sides of the body may be easily contrasted. The number of galvanic cells employed or the number of milliampères of current (as shown by a galvanometer) should also be specified, and the faradaic reaction of homologous nerves or muscle should be stated for the purpose of comparison and for clinical deduction. We may follow with advantage some such plan as the following:

NAME, ...   DATE, ...   AGE, ...

HISTORY OF CASE. See page .... of CASE-BOOK.

FARADAIC TESTS.

	Right side.	Left side.	Extent of secondary coil employed. (In centimetres.)	Nerve tested.
Nerve-reactions.....				
				Muscle tested.
Muscle-reactions				

GALVANIC TESTS.

	Right side.	Contraction produced.	Left side.	Nerve or muscle tested.
	Cells or milli-ampères.		Cells or milli-ampères.	
Nerve-reactions..		C. C. C. A. O. C. A. C. C. C. O. C.		nerve.
Muscle-reactions.		C.C.C. A. C. C. A. O. C. C. O. C.		muscle.

Slips of this character may be printed and kept on hand. They can be pasted into the case-book of the physician when filled out. The tests made at different dates can thus be compared with each other and the progress of each case determined.

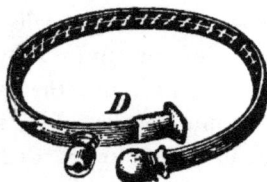

Fig. 34.—THE AUTHOR'S SPRING ELECTRODE.—D, the binding-post for attach-ing the rheophore which connects it with the battery, or with the diagnostic key-board when that instrument is employed. The motor point of the elec-trode is represented as enveloped in chamois-skin. It must be thoroughly dampened in salt-and-water before it is applied to the nerve or muscle to be tested. The other end of the electrode is designed to prevent slipping of the instrument after its proper adjustment.

For the purpose of demonstrating the special action of individual muscles and nerves before classes of students, as well as the study of muscle- and nerve-reactions in disease, I have devised small electrodes which may be made station-ary upon any desired part of the head, limbs, or trunk, by means of straps, strips of adhesive-plaster, or insulated springs. By means of these I have been enabled to make many points clear to a large audience which would be extremely difficult to show by any other method. Furthermore, it is often desirable to refer from time to time during an examination of a patient to the effects of currents of known intensity upon certain nerves and muscles for the sake of accurate comparison, etc. Small electrodes of the type described may be accurately placed upon a patient and allowed to remain upon the spot selected during the entire examination. To each of these a separate rheophore may be attached and, by a simple device of my own, each may be controlled by touching a key upon a board, without movement of the operator. I can thus observe simultaneously the reactions of corresponding muscles or nerves upon the

Fig. 35.—The Author's Diagnostic Key-board. —*A*, the rheophore which connects it with one of the binding-posts of a galvanic battery ; *B*, rheophores connecting its binding-posts with spring electrodes previously placed upon the body of the patient so as to influence the nerves or muscles to be tested ; *C*, buttons and springs which make a circuit to the body of the patient when the knob on the spring is pressed downward so as to impinge upon the button. The number of rheophores which may be employed depends upon the necessities of the case ; the cut shows an instrument capable of six.

two sides, those of the leg and arm of the same side, and any other comparisons which may be required in diagnosis.

The "motor-points" of the body are not always exactly where charts depict them; hence it is sometimes necessary to hunt for them within a radius of an inch or two of the normal point. When they are found with exactness, a small electrode may be fastened over the spot (with moistened absorbent cotton beneath it) and allowed to remain stationary during the entire sitting. Whenever it becomes necessary to refer to the reactions of that point, it can be called into action by touching the key connected with it by its individual rheophore. The cuts introduced show the arrangement of my device for this purpose. I have given a more complete description of the advantages of this method over others previously employed, in the "New York Medical Journal" of May 9, 1885.

Now, from such a table of record it is apparent that the *faradaic current should first be employed* upon the patient (the poles of the secondary coil being used). The extent of the overlap of this coil (in centimetres) necessary to produce muscular contractions when the nerve- and muscle-reactions are being separately tested should be recorded. In case no muscular contractions ensue, the extent of the overlap which produces an *unbearably painful current* should be ascertained and noted. This may be compared with that necessary to produce contractions upon the healthy side. The "polar method" should be employed in the tests thus made and recorded.

The next step in the examination consists in *changing the rheophores to the binding-posts of a galvanic battery.* We can now ascertain the number of cells or of milliampères (which is preferable) required to produce the different varieties of contractions (enumerated in the table designed for record) of muscles in homologous regions of the right and left sides. Each nerve which is impaired should be tested first; and the muscles supplied by it should be tested afterward. The strength of the current employed should be ascertained by throwing a galvanometer into the circuit (when extreme accuracy is desired); by so doing, a compari-

FIG. 36.—THE AUTHOR'S DIAGNOSTIC KEY-BOARD AS APPLIED IN ACTUAL USE. The spring electrodes are represented in the cut (for the purpose of illustration) as applied to the facial, ulnar, and musculo-spiral nerves of each side. If he so chooses, the operator can have his case-book on a stand at his right, for recording his observations as they are made.

son of the nerve- and muscle-reactions of the two sides can be based upon conditions which are exactly alike.

4

When we have completed the steps indicated by the chart prepared for the assistance of the practitioner (page 70) we are in possession of certain facts which may be of great practical value as regards both diagnosis and prognosis:

1. Suppose a case of localized paralysis is examined, and the faradaic and galvanic reactions of both a nerve and its muscles are normal and exactly alike on the two sides. We have reason then to believe that the exciting cause is either hysteria, a lesion of a higher spinal segment than that from which the nerve arises, or a lesion within the brain, provided the possibility of deception on the part of the patient respecting his paralytic condition can be excluded.

2. If the *nerve-reactions* of the affected side to both currents *are exaggerated* (i. e., if the contractions occur in their proper sequence, but under a weaker current than in health), the *probability of an existing central lesion is heightened*, although hysteria may possibly still exist as the exciting cause of the paralysis.

3. If the faradaic current applied through the nerve *fails to produce contractions* of the affected muscles as readily as upon the healthy side (i. e., if a stronger current is demanded to call any one of the paralyzed muscles into action indirectly through the nerve which supplies it), then we know that the *nerve filaments within the spinal cord* or those of the *trunk of the nerve itself* are affected by a lesion which *has impaired but not entirely destroyed their usefulness.*

4. If *no current* from a faradaic machine (which can be endured by the patient) *causes muscular contractions*, we know positively that the *motor cells of the anterior horns* of that spinal segment which controls the paralyzed muscles are destroyed, or that the *nerve itself has been severed* from its connection with the spinal cord.

5. When the MUSCLE-REACTIONS to the faradaic current have been tested, the previous deductions (based on the

nerve-reactions) still hold good. The electrode should, however, be placed over the "motor point" of each muscle thus tested. These are shown in plates at the end of this volume.

6. If the *formulæ obtained by the galvanic current are normal*, all questions regarding the existence of degenerative changes in the nerve- or the muscle-plates can be excluded. When the *normal order is altered*, degenerative changes in the nerve- or the motor-cells of the spinal cord are present.

7. The history of a case in which motility is impaired is never complete without a record of an electrical examination of the nerve- and muscle-reactions to both the faradaic and galvanic current. When doubt exists respecting the existence of a cerebral lesion or hysteria, the facts obtained by other methods of examination (fully described by me in the "Medical Record," March, April, and June, 1884) will clear up all doubts.

8. Patients afflicted with *paralysis from a cerebral lesion* generally exhibit normal electro-nerve and electro-muscular reactions in the paralyzed parts. In some instances the reactions may even be exaggerated.

9. *Hysterical patients* afflicted with paralysis may exhibit either normal or exaggerated electro-muscular reactions to faradism or galvanism. The sensitiveness of the muscles to faradism is generally decreased; in some cases it may be totally wanting (Duchenne).

10. In *rheumatic paralysis* the electro-muscular contractility is, as a rule, markedly increased; this may be shown by a comparison of the reactions of the two sides of the body. In exceptional cases this is not found to be so, as I have seen the reactions follow only the strongest currents.

11. In *peripheral paralysis* the faradaic and galvanic reactions are altered after ten days have elapsed. The mus-

cular contractility to the faradaic current is lost early to a greater or less extent; and the formula of degenerative changes is developed later by the employment of the galvanic current.

12. A decreased musculo-excitability to the faradaic current in the musculo-spiral nerve and the extensor muscles of the forearm on both sides—the flexors being normal and the lower extremities being unaffected—tells us of the approach of *lead-poisoning* before the actual symptoms are well marked.

13. In *progressive muscular atrophy* a response to the faradaic current can be obtained as long as any fibers in the muscle tested remain free from fatty metamorphosis.

14. *No alterations* in the electro-contractility of muscles is observed in any of the diseases confined to the *posterior part of the spinal cord.*

DETECTION OF FEIGNED DISEASES.

In addition to the uses of electricity for the purpose of determining the presence or absence of nerve- and muscle-degeneration, and the discrimination between cerebral and spinal paralysis and the various types of peripheral palsies, some other useful purposes in diagnosis have been published from time to time.

The Detection of Death.—It is stated that muscular contractions produced by the faradaic current can not be maintained over four hours in a dead subject.

Malingerers are not uncommon among the applicants for charitable aid, and they are occasionally encountered among the higher walks of life. Epilepsy and paralysis are the most common diseases which are feigned.

Feigned epilepsy can be distinguished by the application of a strong faradaic current to the forehead or tibia by means of a wire-brush. The intense pain so produced will not be appreciated by a true epileptic, but will bring the fit to a sudden close, if assumed in order to create sympathy or aid.

Feigned motor paralysis is usually exhibited in some of the various forms of peripheral paralysis. Few malingerers know enough to simulate hemiplegia or paraplegia without detection. If two weeks have elapsed since the attack, the presence of normal electrical reactions of nerve and muscle in the affected limb is strong ground for suspicion, provided a history of some cerebral lesion or of hysteria can not be elicited. There are various other tests which a skilled anatomist can employ in each case that will help to clear up all doubts upon this subject.

Feigned anæsthesia may be told by the use of the faradaic current with the wire-brush attached to the negative rheophore. The opposed limbs will quickly show how much actual anæsthesia exists.

DETECTION OF BULLETS OR BURIED METAL.—An ingenious application of electricity to surgery has been made which has utility in diagnosis.

The so-called "*electric probe*" consists of two wires which are perfectly insulated from each other by rubber or ivory. These wires terminate in metal tips which project slightly beyond the insulating medium, and at the other end of the wires a galvanic cell and an electro-magnet acting upon a bell are attached. When the probe is pushed into the tract made by a metallic missile, and the tips are brought into contact with it, the circuit is completed and the bell rings. The animal tissues are not sufficiently good conductors of electricity to form a circuit; hence the bell will not ring until the metal is touched by the tips of the probe.

ELECTRO-DIAGNOSIS OF AURAL DISEASES.—We owe to Brenner, of St. Petersburg, the first suggestions of this use for electric currents in diagnosis. From a somewhat limited experience in its use, I am led to believe that its utility can not longer be called into question. Brenner's formula seems, furthermore, to be in accord with all that has been proved in respect to nerve-trunks in health, in all parts of the body. The formula is simply that of the galvanic reactions of the auditory nerve in health.

1. In place of the C. C. C. observed in muscle, we get, when a galvanic current is sent through the auditory nerve, a *ringing noise* when the *cathodal closure* occurs. (C. C. S.)

2. The *cathodal opening* produces no effect.

3. The *anodal opening produces a ringing noise* when a current of high intensity is employed.

4. After the cathodal closure (*cathodal duration*—C. D.), the ringing noise produced at the closure *gradually diminishes.*

The formula which is indicative of health when a pole is connected with each ear may be expressed in symbols, as follows:

	RIGHT EAR (Anode).	LEFT EAR (Cathode).
C. (closure)....	....	S. (loud)
O. (opening)....	S. (weak)	......
D. (duration)...	....	S. >

Now, it is maintained by Brenner that any deviation from the normal reactions of the auditory nerve (shown in the preceding formula) indicates disease of the acoustic mechanism. The variations produced by the different diseased conditions encountered can not be given here from want of space.

In applying galvanic currents to the ear, it is best to place a medium-sized electrode over the entire tragus or to fill the external auditory canal of the ear to be tested with tepid water containing a little salt, and then to introduce an electrode of metal inclosed in an aural speculum of hard rubber into the ear until the metal touches the water. If each ear is to be tested separately, the other electrode should be placed at an indifferent point, preferably the mastoid region of the same side or the middle of the sternum.

Regarding this test it is well to state that repeated sittings are often necessary. The patient has to become accustomed to the disagreeable effects of the current. It is desirable that you begin with very weak currents and increase the strength very gradually. As the reaction of *cathodal closure* (C. C.) is the most important, it can be intensified by previously allowing cathodal duration (C. D.) to act, or by rapidly following C. C. by A. O.

EXAMINATION OF THE EYE, NOSE, AND MOUTH BY ELECTRICITY.—Special effects are produced upon the organs of sight, smell, and taste by electric currents. Sparks or vivid flashes of light follow electric stimulation of the eye; and subjective odors and tastes are also produced when the olfactory and gustatory nerves are excited by this agent. The "polar method" of examination is employed when any of the special senses are thus tested. I would caution each of you against employing too strong currents about any of the organs of special sense. The examination of the optic, olfactory, and gustatory nerves requires experience and should not be attempted by novices.

When examining the *reactions of the eye*, the small electrode is placed upon the closed eyelid, temple, or forehead. The large electrode is placed upon the back of the neck. The room should be darkened and the patient should keep both eyes closed.

When *testing the sense of taste*, the poles should be in contact with the cheeks, and the sensations of taste experienced upon both sides by the patient should be ascertained. A fine electrode can also be placed upon the tongue, the pharynx, or the inside of the cheek, in case localized polar reactions are to be determined. A double electrode, with two metal tips which are not in contact, may be employed for this purpose (Neumann).

ELECTRICAL EXAMINATION OF THE SENSIBILITY OF THE SKIN.—The electrode devised by Erb is to my mind the best for this purpose. It consists of four hundred varnished wires in a tube of hard rubber. The ends of these wires make a perfectly smooth surface. This electrode is connected with the secondary coil of a faradaic machine and is then pressed upon the area of the body to be tested—the other pole being at the sternum. The *minimum* of the overlap of the secondary coil which can be felt and the *maximum* which can be endured are both recorded. Homologous parts of each side should be compared with each other.

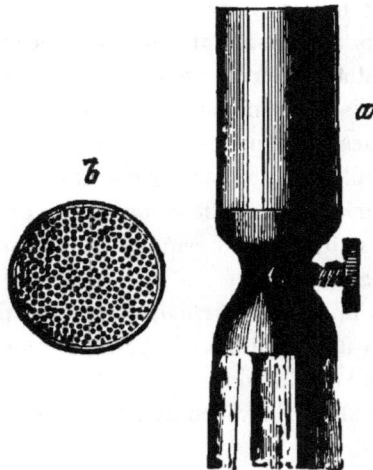

FIG. 37.—ERB'S ELECTRODE FOR THE EXAMINATION OF FARADO-CUTANEOUS SENSIBILITY. *a*, a hard-rubber tube ; *b*, the free surface of the electrode.

Regarding this test Erb wisely remarks : " The skin, regarded as a sensory organ, can not be tested with irritants other than those adequate to it—viz., touch, pressure, various temperatures, and the higher grades of those irritants which produce pain. It may be disputed whether electricity should be included among these 'adequate' irritants of the skin. The electric sensation is a specific, distinct quality of tegumentary sensibility, whose careful examination, however, is of value in many morbid conditions."

Part III.

ELECTRO-THERAPEUTICS.

WE have now discussed the essential points pertaining to electro-physics and electro-diagnosis, and there remains now for us to consider the uses of electricity in the treatment of diseased conditions of various organs and tissues.

Before we pass to details of the practical part it may be well for us to review in a general way some of the laws which should govern us in applying electric currents to the different tissues, and the objects to be attained by the employment of faradaism, galvanism, and static electricity.

GENERAL ELECTRO-THERAPEUTICS.

The rapidity and completeness of reported cures by the use of electric currents upon living tissues during the last quarter of a century leave no room for doubt that this agent is particularly valuable in the treatment of paralysis, neuralgia, spasmodic diseases, disturbances in the sensibility of the skin, and many disordered states of the brain, spinal cord, and peripheral nerves themselves. We have undisputed facts which prove also that blood may be coagulated with safety within some aneurysmal sacs by the galvanic current, that the life of the ovum may be destroyed in extra-uterine pregnancy, that animal tissues may be disintegrated by chemical changes induced within them by this agent, and that neoplasms may be removed without hæmorrhage by the cautery loop.

Our present ignorance of the molecular and nutritive changes in tissues (as the cause or result of disease) renders it impossible to do more than speculate upon the theory of the therapeutical action of electricity in many cases; but, on the other hand, our empirical knowledge of these effects is none the less valuable because we are unable to explain them. The same criticisms would otherwise hold good in reference to almost all of the drugs employed in medicine. None of us know exactly *how* they produce their specific effects.

Concerning speculation upon electrical effects on living tissues, Erb remarks as follows: " What appears more natural than that neuralgia and spasms could be relieved by the sedative action of the anode, with production of anelectrotonus, and that anæsthesia and paralysis could be cured by the exciting action of the cathode, with production of catelectrotonus? But, apart from the fact that we are not certain that an increase of irritability really occurs in one group of cases and a diminution in the other, it must be remembered that electrotonic action disappears very rapidly after the cessation of the current, while the curative effects of the current are more or less permanent."

Now, we may summarize the general principles which regulate the use of electric currents as follows:

1. They may exert, under certain circumstances, a *stimulating* or *irritating effect*. This is, perhaps, the basis of the most varied applications of electricity to disease.

2. They may exert, when properly applied, a *sedative action* on nerves or nerve-centers.

3. They may be made to exert a *catalytic action* upon neoplasms, enlarged glands, etc.

4. They are capable of causing *electrolysis.* This action is one which has lately come into prominence.

5. They create *heat* under certain conditions. The galvano-cautery is to-day assuming a very prominent place in some of the departments of surgery.

Let us now discuss each of these special actions separately, noting the general points of interest pertaining to each which will aid us in properly treating our patients. Electrolysis and the galvano-cautery have been treated of in previous lectures.

STIMULATING OR IRRITATING EFFECT OF ELECTRICITY.— This is indicated in many diseased conditions. Among these the following may be prominently mentioned:

Some of the various forms of cerebral and spinal diseases.

Depressed irritability of some special nerve-trunks.

Abnormal resistance to conduction of electric currents, exhibited by the motor or sensory nerve-filaments of some part.

As a counter-irritant to some pathological conditions.

Trophic disturbances of special regions (skin, nails, hair, etc.).

Vaso-motor depression.

Atrophic changes in muscles.

As a means of indirectly affecting the nerve-centers through the sensory nerves, thus influencing respiration, circulation, phonation, vaso-motor paths, peripheral organs, the muscles, etc.

The methods of application which are best adapted to accomplish irritating or stimulating effects are differently stated by authors. Personally, I do not confine myself exclusively to faradaism or galvanism,

The *faradaic current* is more commonly employed for this purpose than the galvanic. The electrodes should be selected, as to their size and shape, in accordance with the parts to be acted upon; they should be well moistened with salt water, and kept closely in contact with the skin. The wire-brush is the best electrode to stimulate the nerves or other tissues of the skin. It should be used dry. I prefer the secondary faradaic current to that of the primary coil for stimulating effects.

If galvanism is employed as a stimulant, Remak's plan, of moving the well-moistened cathode rapidly over the nerve-trunk or muscle to be stimulated, with a current sufficiently strong to cause strong wave-like contractions, is a good one. Another method, termed by this author "terminal labile stimulation," consists in stroking the tendinous end of a muscle with the cathode so as to affect the entire length of the muscle. In both of these methods the anode is kept stationary upon some indifferent or neutral point—the center of the sternum by preference, or the nape of the neck.

One of the most vigorous methods of stimulation consists in rapidly changing the polarity by means of a commutator, when the galvanic battery is employed.

The Combined Current.—Another method which I employ (not generally mentioned in text-books) consists in connecting a galvanic battery, by means of a rheophore, with a faradaic instrument, thus bringing *both a constant and induced current* to bear upon the tissues at once. The rheophore which connects the batteries joins the positive binding-post of the galvanic instrument with the secondary coil of the faradaic; the two rheophores connected with the electrodes run from the negative binding-post of the galvanic and from the secondary coil of the faradaic instrument. The two instruments (faradaic and galvanic) are thrown into action simultaneously, and the strength of the current employed is graduated by the number of cells used in the galvanic battery and by the extent of the overlap of the secondary coil of the faradaic instrument. I have obtained some remarkable results by the stimulation thus produced in various forms of trophic disturbances of the skin and muscles.

The stimulation of nerve-fibers (when obstacles exist to their conduction) should be performed peripherally from the site of the lesion in sensory nerves, and as centrally as possible in motor nerves (Erb). Degenerated and atrophied nerves and muscles require a direct effect of the currents employed. For these reasons, the site of stimulating electrical applications depends upon the situation and character of the lesion and the object to be attained.

MODIFYING EFFECTS OF ELECTRIC CURRENTS.—The *irritability of nerves and muscles* may be influenced by electric currents.

In certain diseased conditions we may expect a favorable result from such an action. Thus, for example, in some types of paralysis, in anæsthesia, in certain vaso-motor disturbances, and in depressed states of cerebral and spinal activity, the irritability of nerves or of muscular fibers is diminished; hence we resort to the so-called " catelectrotonic action " of electricity as a means of stimulating and restoring the normal irritability of the tissues affected.

It is now generally accepted as proved that *feeble faradaic currents* will accomplish this end. Galvanic currents, when applied for this purpose, give more positive results, however, than faradaic.

In order to *increase irritability* by galvanism, the negative electrode should be applied in a stabile manner (*i. e.*, without being moved) to the part upon which this effect is to be produced; and the strength and duration of the current should be steadily increased. When the muscles or motor nerves have been exhausted by over-exertion, excessive fatigue, etc., this action (termed by Heidenhain the "refreshing action" of galvanism) is particularly indicated.

Those conditions in which the *normal irritability of nerves or muscles is intensified* demand the so-called "anelectrotonic action" of electricity. These conditions comprise all irritative states of the sensory, motor, and vasomotor tracts within or without the brain and spinal cord; hence we employ this action in neuralgias, spasmodic affections, hyperæsthesia of any of the cerebro-spinal nerves, headache, excitation of any of the special senses, cerebral and spinal irritation, etc.

In order to *decrease the irritability of nerves or muscles,* we may employ very powerful faradaic currents. We may also begin by employing a feeble faradaic current and gradually increasing its strength to the highest point of endurance; then maintaining it at this point for some time; and subsequently reducing it gradually to the feeblest current perceptible to the patient. This method is known as the "*increasing induction method.*" Electrodes well-moistened and of large size should be employed and kept immovable upon the same points during the application. It is often advisable to repeat this procedure several times at one sitting (Erb).

When the galvanic current is employed for the purpose of decreasing irritability, the positive pole is made fast at the point to be influenced. The current is increased in strength and maintained at its maximum for some time, after which it should be decreased gradually until it can not be perceived by the patient. The gradual decrease of the current-strength prevents the marked temporary increase of irritability which is liable to follow this method when this step is omitted.

Static electricity exerts in many cases an immediate beneficial effect upon neuralgic pains (especially upon sciatica)
and upon the various spasmodic affections, as, for example,
chorea, paralysis agitans, tremor, contracture, etc. These
effects are obtained, in some cases, when galvanism and
faradaism have proved of no benefit. I should never regard
a case as incapable of benefit by electric treatment until
static electricity, in the form of insulation, the electric wind,
or the spark, had been thoroughly tested. I have had better results with this form of current in tremor than with
galvanism or faradaism.

Some forms of pain (as, for example, the pains of ataxia,
sciatica, trigeminal neuralgia, muscular rheumatism, etc.)
are oftentimes relieved by a few applications of static electricity. My experience with this agent has convinced me
that its effects are often satisfactory in cases where pain is a
prominent symptom, when galvanism has been tried without benefit. I have found that insulation and the abstrac-
tion of heavy sparks from the seat of pain give the best
results.

CATALYTIC ACTION OF ELECTRICAL CURRENTS.—Under
this heading we include (1) an increase of absorption produced by dilatation of the capillary blood-vessels and lym-
phatics; (2) an increased capability of tissues for imbibi-
tion of fluids, through an increase of osmotic processes;
(3) changes in the disassimilation and nutrition of nerves,
on account of their stimulation or "refreshing effects; (4)
changes in the molecular arrangement of tissues, caused
by electrolytic processes; and (5) the results of the transportation of fluids from one pole to the other (Remak
and Erb).

Remak has shown that muscles become congested and
greatly swollen when subjected to galvanism. They are
rendered tense, and (according to this observer) absorb

water more freely than muscle which has not been galvanized.

Changes of a marked character may be induced in the skin by galvanism. These have been studied by Erb, Remak, Bollinger, and others.

The vaso-motor nerves may be influenced by electrical currents. This is shown by many of the later investigations—prominently those of Lowenfeld, which apparently demonstrate that contraction and dilatation of the vessels of the brain result, respectively, from antero-posterior and transverse currents through the head from a galvanic battery.

Although we are, as yet, unable to speak with positiveness regarding the certainty of the catalytic effects of electrical currents, or to map out the forms of disease which are to be regarded as specially indicating these catalytic effects, still it may be said that the following states have been successfully treated by electrical currents, and that the cures are probably to be attributed to a catalytic action: (1) Inflammatory affections of the nervous system, including sclerosis, myelitis, neuritis, etc.; (2) arthritis and chronic exudations into joints; (3) glandular enlargements; (4) hard cicatrices, periosteal swellings, and fibrous adhesions; (5) contusions, sprains, extravasations of blood, and other results of traumatisms.

The *galvanic current* is the one that is generally employed when catalytic effects are desired. In diseased conditions of the brain, spinal cord, or any of the deeply seated organs, the faradaic currents are not usually productive of benefit.

The "stabile method" of application of the galvanic current is preferable, to my mind, when catalytic action is to be attained. The strength of the current should be sufficient to easily overcome the resistance offered, and the duration should be sufficiently prolonged to accomplish changes in the tissues subjected to its influence. One pole is placed, as a rule, at an indifferent point (the sternum by preference), and the other over the tissue diseased. Sometimes, as in the case of the brain, for example, the poles are placed upon either side of the diseased part. Although there are exceptions to the rule, it is well to use the anode or positive pole over the diseased part when pain is present, when symptoms of active irritation exist, or when the morbid processes are very active. The cathode or negative pole is best adapted to influence chronic morbid processes, such as sclerosis, indurations, etc. Erb recommends that the polarity of the current be changed several times in either case; he doubts the infallibility of the rule given, although it is theoretically sound. Chvostek urges the use of short and moderate currents for a few minutes (three to ten) when catalytic action is desired. In this way, he believes, the vaso-motor and trophic nerves are more impressed than by any other method.

Respecting the catalytic action of *faradaic currents*, a difference of opinion exists between authors of note. One thing is certain—viz. that strong currents are required, and that the currents must be passed directly through the diseased part to accomplish marked results. Glandular tumors have been resolved by this method with great rapidity in some recorded instances.

GALVANIZATION OF THE CERVICAL SYMPATHETIC.—This method has afforded relief, according to published cases, in vaso-motor and trophic disturbances of the nerve-centers, the eye, viscera, muscles, joints, and skin. Thus, for example, cases of cure of epilepsy, atrophy of the optic nerve, Basedow's disease, progressive muscular atrophy, lead-palsy, scleroderma, chronic rheumatic arthritis, bulbar paralysis, neuralgias of various types, and many other conditions have been reported by means of this method. Respecting this step, Erb wisely remarks as follows: " There can probably be no doubt of the correctness of a part of these observations, but this does not by any means imply that the cervical sympathetic is responsible for such results."

When we review the structures which compose the neck and recall the numerous connections which exist between the sympathetic cords, the pneumogastric nerve, the brain and cervical segments of the cord, the medulla oblongata, the brachial and cervical plexuses of nerves, etc., it becomes clear why De Watteville applies the term " sub-aural galvanization " and Erb the term " galvanization of the neck " to this special procedure.

The steps required to influence these parts by Meyer's method consist in the application of a small electrode (cathode) under the angle of the jaw and adjacent to the hyoid bone, and then crowding it backward and upward against the vertebral column, the positive electrode (of larger size) being placed over the seventh cervical spine. The current may be stabile, labile, or interrupted; or the polarity may be changed from time to time during the sitting of from one to three minutes. Six to ten galvanic cells of the Grenet variety are sufficient. The application may be unilateral or bilateral, according to the demands of the case.

Corning has devised an instrument which insures carotid compression with galvanization of the neck for the treatment of cerebral hyperæmia and some other morbid conditions.

Benedict places the positive pole in the jugular fossa, and the negative pole upon the superior cervical ganglion.

GENERAL FARADIZATION.—This method of administer ing electricity was first employed by Beard and Rockwell. By this procedure the entire body is subjected to secondary faradaic currents of varying intensity. It is applicable chiefly to those forms of nervous disturbance which are associated with general debility, poverty of the blood, special diatheses and cachexiæ, hysterical affections, skin diseases, persistent chronic inflammations, and other results of low vitality or functional derangements of the organs.

To apply this method, the patient must be undressed or very loosely clothed. The feet are immersed in a bowl of tepid water with a little salt added, in which the cathode is also placed after being connected by means of a rheophore to the binding-post of the secondary coil of a faradaic machine. The anode is held in one hand of the physician, and his other hand (well moistened in salt water) is applied to all parts of the surface of the patient's body. If the subject can bear it, a large electrode covered with absorbent cotton and flannel, or with a soft sponge, is employed in place of the hand. The application should begin at the head and terminate at the feet—the strength of the current being modified from time to time as the feelings of the patient may demand. The extremities and back should have vigorous stimulation, the nerves of the neck should be influenced by a much weaker current, and the cœliac plexus should be influenced by a stabile application of a few minutes over the epigastrium. The entire duration of the application occupies from ten to twenty-five minutes. It may be applied as often as three times a week if necessary.

Personally, I can attest the efficacy of this treatment as a general tonic. I have witnessed immediate effects from it in some of my cases, and I employ it constantly in a modified form.

In case the hand of the attendant is to be employed as an electrode, I would advise you to use the instrument which I now show you as a great improvement over the way described by the inventors of this method. It is called the electric bracelet. It is placed upon the right wrist of the attendant over a pad of wet absorbent cotton, and the

rheophore is screwed into the binding-post upon it. The
right hand is then wet in salt water and used as previously
described. By this instrument the hand of the attendant
is alone subjected to the current, and the fingers can detect
muscular contraction in the patient even when too feeble to
be seen easily. As an adjunct to massage, I employ this
useful instrument with decided benefit.

GENERAL GALVANIZATION.—The steps required by this
method are similar to those previously described, except
that the constant-current battery is employed in place of a
faradaic machine.

CENTRAL GALVANIZATION.—The cathode is placed over
the epigastrium. This electrode should be of large size.
The anode is stroked over the forehead, with a current of
about two milliampères, for two minutes; then made stabile
over the cranium for about two minutes; then moved up
and down the neck on each side for the same duration;
finally it should be moved along the length of the spine for
about five minutes. This method was a favorite one with
the late Dr. Beard, who reported cures of gastralgia, hysteria, hypochondriasis, nervous dyspepsia, and many of the
symptoms of cerebral and spinal neurasthenia by its continued use. In two cases of gastralgia in which I personally employed it for some time I obtained an absolute
recovery.

THE ELECTRICAL BATH.—This method of administration
of electricity to a patient may be accomplished by using a
metal tub, or one which is composed of a non-conductor. If
a metal tub is employed, the patient must be protected
from actual contact with it. This is usually accomplished
by means of wooden slats or some other medium of support
for the patient when immersed. If the bath-tub is of
metal, one rheophore of the battery employed is attached
to the tub, while the other is attached to an electrode held

by the patient or placed in contact with his body. If the tub is of a non-conducting material, both electrodes may be placed in the water. The electrodes employed should be very large (often running the entire length of the tub), in order to allow of as great a diffusion of the electricity as possible. The fluid in the tub may be simple water, or, preferable, a solution of salt, soda, or an acid.

Personally, I am not a strong advocate of this method of treatment. It violates one of the fundamental principles of electrical treatment of localized affections in that it does not confine the polar action to the part or parts diseased. In the second place, I have not found its tonic action to equal that of general faradization or general galvanization.

Strong claims have been made in its favor as a remedy for tremor (especially of the alcoholic and mercurial varieties) and for chronic articular rheumatism, but I am not yet convinced that they are to be regarded as well-established.

If you desire to try this method of treatment in any case, it is well to know that the temperature of the bath, as well as the strength of the current employed, should be modified by the condition of your patient. The duration of the bath should never exceed thirty minutes, and ten minutes will generally suffice. The current should be strong enough to be perceived by the patient in all cases.

THE RELIEF OF PAINFUL POINTS.—One of the most generally useful effects of electricity is the relief which it affords in many cases to pain. Of all the methods of treatment of neuralgia now employed, I consider electricity, in some of its various forms of application, by far the most efficacious. Personally, I have almost discarded internal medication for the relief of this class of sufferers. You have personally witnessed in one or two cases brought before you the immediate relief which follows the application

of electricity to painful points, and you have seen neuralgia
cured at one sitting by the same agent.   Do not understand
me, however, as maintaining that this can be accomplished
in all cases.   Repeated applications are demanded, as a rule,
and, in some instances, months of treatment are required to
insure a cure of this distressing malady.

In the majority of subjects afflicted with neuralgia, painful points may be detected along the course of the affected
nerve or its branches.   These are situated, as a rule, where
the nerve gives off a branch or bifurcates, and also where it
passes through a foramen.   Sometimes it is necessary to
make pressure along the course of the nerve to detect the
existence and seat of these points.

Now, it should be remembered by each of you that the
successful treatment of neuralgia depends largely in some
cases upon the *direct treatment of these painful points.*   They
seem in some way to have a relationship with both the production and cure of neuralgia, as well as spasm of the muscles, ataxic symptoms, and other forms of nervous diseases.
These points may be the seat of a localized periostitis, a
circumscribed inflammatory exudation, a neuritis, an enlarged gland, and many other conditions which create nervous phenomena.   In a few instances the symptoms even
of ataxia have been relieved, by the electrical treatment of
painful points in the region of the spinous and transverse
processes of the vertebræ, by men of note, among whom
may be mentioned Brenner, Remak, Meyer, Legros, and
others.

The steps which you should employ in the treatment of
painful points are as follows : 1. Use the galvanic current,
employing from three to eight Grenet cells. 2. Apply the
anode to the painful spot, and keep it stationary at that
point. 3. Place the cathode at some indifferent point, preferably the sternum. 4. Do not use a current which will be

excessively painful to the patient, nor exceed five minutes in the application. I frequently do not allow the duration of the current to exceed two minutes at a sitting. It is advisable, in persistent cases, to make the applications daily.

Of late some experiments have been made, with apparent benefit, by having patients of this class wear over the painful points *a piece of metal*, connected with another piece of metal (which is also in contact with the skin) by an insulated wire. The best metals are zinc and copper. They should be brightly polished before the application, and should have a piece of dampened linen between them and the skin. They may be worn continuously for weeks, or changed each day on retiring and rising.

Some authors recommend the employment of *very feeble galvanic currents for an hour or two at each sitting,* the anode being placed over the painful point. Le Fort goes so far as to suggest the propriety of applying such currents continuously for weeks, by means of ordinary rheophores and electrodes, when fatty changes, contractures, or reflex paralyses are to be combated.

ELECTROLYSIS. — When a galvanic current is concentrated within animal tissue by a close approximation of the electrodes, or when, by means of the "polar method" and insulated needles, a galvanic current of high intensity is made to traverse some selected spot upon the human body, there is apt to be a chemical decomposition of the water and salts, and a coagulation of the albuminous elements of the tissue thus acted upon. The salts are then separated into their bases and acids, while the water is simultaneously decomposed into hydrogen and oxygen. The positive pole attracts to it the acids and the oxygen; the negative pole attracts to it the alkalies and the hydrogen. For this reason the insulated needles tend to become oxidized when they are connected with the positive rheophore. They do not become so when attached to the negative rheophore. The free alkalies deposited at the negative electrode are apt, on the other hand, to cause destructive effects upon adjacent tissues. These are greatly in excess of that produced by the oxidation of the metal points of the insulated needles when joined to the positive rheophore.

When we wish to test the strength of the current which we propose to employ for electrolysis, it may be easily done by sending the current through the white of an egg for twenty or thirty minutes. In that time it should coagulate the albumin.

In order to reach the parts upon which we most desire to perform electrolysis, it is often necessary to perforate the skin and the muscles. To do this, needles are employed. They should be insulated with hard rubber, collodion, or shellac, except at their point for one half-inch, and the uncovered part should be gilded, as a rule, in order to prevent its oxidation. They should be from two to five inches long; should be strong enough to penetrate tissues without a liability of breaking; should be as small as is consistent with the current-strength to be employed; and should be so arranged as to enable the operator to adjust them in a handle to which one of the rheophores of the battery may be attached. Ordinary sewing-needles strung on a wire may be employed in treating superficial nævi, tumors, etc., if you lack the instruments specially designed for the purpose. You may shellac them for insulation if deemed best.

It is very important, in some cases, that the insulation of the needles employed be as perfect as possible; and that the tips of the needles be triangular or lancet-shaped, in order that they may penetrate the skin with ease. The needles and handle required can be bought of any manufacturer of electrical appliances.

The battery employed for electrolysis need not be un-

like that for ordinary medical purposes. Twenty-four of
Grenet's cells will produce a sufficient intensity of current,
provided they are freshly filled. Robin's statement that a
current of forty-five milliampères is requisite must be based
upon a very limited external resistance. It is well to use a
battery of greater power than is actually required, so that
fresh cells can be added, without breaking the current during the operation, when deemed necessary. Personally, I
prefer insulated copper wire for rheophores when performing electrolysis, in preference to the tinsel cords commonly
employed for electric applications. They are less flexible
than the tinsel cords, but they are vastly superior to them as
conductors.

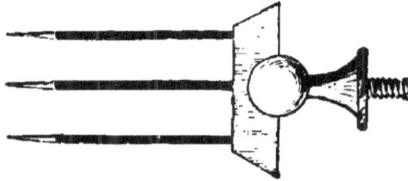

FIG. 38.—ELECTRODE FOR ELECTROLYSIS (with three insulated needles).

Electrolysis has been employed for the following purposes, with more or less success :

   1. The coagulation of blood in aneurysmal sacs.

   2. The relief of nævi and erectile tumors.

   3. The cure of cystic tumors.

   4. The cure of goitre.

   5. The cure of echinococci.

   6. The cure of ovarian cysts, and those of the broad
ligament.

   7. The cure of urethral stricture.

   8. The cure of malignant and fibroid growths.

   9. The cure of hydrocele.

   10. The destruction of the fœtus in extra-uterine pregnancy.

5

11. The removal of cicatrices, polypi, and other new formations.

Respecting the treatment of large aneurysmal sacs by this method, statistics show that the operation merits more general recognition, as a means of possible cure of intra-thoracic and abdominal aneurysms which can not be safely treated by ligation, than it has received. Nine cures out of thirty-seven cases have been collected by Duncan from various sources; and Bartholow has since collected others which have been benefited by it, although not positively cured. In none of these cases, so far as I can ascertain, was the current-strength measured by a galvanometer. There is reason to hope that the operation may become more generally employed when the steps of the procedure necessary to its success are determined with greater exactness. There seems to be a doubt, as yet, regarding the best method both of generating and directing the current, so as to prevent suppuration and secondary hæmorrhage. In two cases of blood-sacs upon the face, intrusted to my care, I have succeeded in consolidating the tumor and effecting its radical cure by this means without suppuration or other complications. I employed the positive rheophore for the needles, and placed the cathode at the nape of the neck in both cases. The duration of the two sittings in each case was about ten

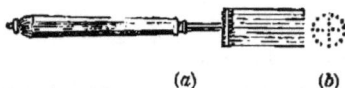

(a)   (b)

Fig. 39.—An Electrode for Electrolysis. The needles are fine and numerous, and are arranged as shown in (b). This is an excellent device for effecting the cure of diffused nævi, small glandular tumors, etc.

minutes, and twenty-four Grenet cells were employed. I believe that the anode produces the firmest clot; hence the danger of hæmorrhage on the withdrawal of the needles is less than when the cathode is employed. The risk of embolism, as a result of disintegration of the clot, does not seem to be so great as one might at first imagine.

The employment of electrolysis in cystic tumors of the neck, the ovaries, and the thyroid gland, has been resorted to by many experimenters of note. Among these may be mentioned Althaus, Amussat, Ultzmann, Clemens, Semeleder, and others. Some of the cases reported seem to point toward this method of treatment of these diseased conditions as potent and comparatively devoid of danger if properly carried out.

The *treatment of urethral stricture* by electrolysis I have always heretofore combated—chiefly because I think it less safe and far less certain than gradual dilatation. I must confess, however, that I have had no personal experience in the electrical treatment of stricture ; and that my prejudices may be more or less without foundation. My perusal of the reported cures by this method has not, as yet, carried to my mind a conviction of my error. The method seems to me to lack precision, which should, to my mind, form the basis of all surgical procedures within that canal. I am having made for me some modifications of urethral electrodes which I believe will insure greater precision than any yet devised ; and I propose to give them a thorough trial.

Ordinary *cases of goitre*, and the *enlargement of the thyroid gland which accompanies Basedow's disease*, have been cured by electrolysis. Rockwell and Butler have reported some astonishing results in the treatment of exophthalmic goitre by galvanism of the thyroid gland. Rockwell places the cathode over that body and the anode over the solar plexus, combined with the employment of the anode in the auriculo-maxillary fossa and the cathode over the cilio-spinal center (cervical segments of the spinal cord) at each sitting. His cases of reported cure required from fifty to sixty-nine sittings. Needles were occasionally employed upon the goître. This treatment was supplemented by the use of iron, zinc, digitalis, and ergot ; and a restricted diet, with instructions regarding the necessity of the repression of the emotions and passions, was enforced.

The *arrest of extra-uterine pregnancy* by electrolysis, and also by shocks transmitted through the sac from Leyden jars charged with static electricity, constitutes, perhaps, one of the most successful and remarkable contributions to medicine. The former method has proved the most reliable. Needles are inserted into the sac and a current is sent through the fœtus of sufficient intensity to destroy life.

The *treatment of cancer* by electrolysis has been followed by satisfactory results in some cases, according to the observations of Beard, Butler, Mussey, and Neftel. The question of accurate diagnosis of the cases reported as cured must still be considered as unsettled. The results apparently obtained should certainly awaken the profession to a trial of this method of treatment of a malady which internal medication, caustics, and the knife seem powerless to combat.

Bartholow reports a cure of four out of six cases of *fibroid tumors of the breast* by electrolysis. The remaining two patients failed to continue treatment for a sufficient length of time. He did not employ needles in any of these cases.

The *relief of hydrocele* by the introduction of two needles connected with the rheophores of a galvanic battery and brought within a half-inch of each other at their points has been reported by Rodolfi, Frank, Bartholow, and others. Some of the cases reported as cured required only one application.

In dismissing this subject it may be well to summarize the effects of electrolysis, as follows:

1. A feeble current tends to cause dilatation of the capillaries and the lymphatic vessels, and thus to aid in absorption.

2. A stronger current decomposes the salts and the water of tissues, and coagulates the albuminoid elements.

3. A disintegration of the tissues immediately adjacent to the pole which produces the effects previously described takes place, with an escape of bubbles of gas, when the decomposition of tissues is active.

4. As an eschar may be formed by a current of great intensity, it is maintained by some authors that the cicatrix which results from such a slough is soft and pliable if the eschar has been made by the anode, and dense, with a tendency to contract, when due to cathodal action. I am unable to confirm or deny this statement.

5. The danger in electrolysis is that of "doing too much" rather than too little. The former error can not be repaired; the latter can by repeated sittings.

6. When an escharotic effect is desired, it is well to have the needles made of zinc. The decomposition of the chloride salts forms indirectly the chloride of zinc, because the liberated chlorine attacks the needle. This is absorbed by the tissues adjacent to the needle, and an escharotic effect is thus produced. In the treatment of malignant growths such needles, with currents of weak intensity, and long sittings, seem particularly well adapted. This method is almost painless, and has produced excellent results in some cases reported.

7. The introduction of needles into the tissues is not an absolute necessity when treatment by electrolysis is indicated. The same effects to a lesser degree may be obtained

by placing the electrodes in contact with cutaneous or mucous surfaces.

8. The employment of iron needles has been suggested for the rapid coagulation of blood, on account of the styptic effect of the chloride of iron which tends to form by the liberation of chlorine from the chlorides of sodium, potassium, and calcium.

THE GALVANO-CAUTERY.—When a large quantity of electricity is forced through the resistance offered to its passage by a platinum wire or a strip of platinum (usually bent into the form of a knife), the heat produced causes the platinum to rapidly approach redness or whiteness. Such an arrangement is known as a "cautery-loop" or a "cautery-knife." The battery which is employed to generate electricity in sufficient quantity to accomplish such a result is known as a "cautery-battery." In cautery-batteries the plates are large and near together; hence unusual precautions have to be taken to prevent "polarization," which takes place very rapidly on account of decomposition of the fluid in which the elements are immersed.

Of all the devices which have been suggested to overcome this difficulty, I prefer that of Dr. Piffard. In the battery devised by him the zinc plates are perforated, so that the fluid can be forced through them upon the platinum plates by means of a rocking motion when the battery is in action. The assistant who operates the battery can produce any degree of heat required by making the plates move slowly or rapidly through the fluid. The key-board of the battery is so connected by means of large thumbscrews that the elements can be connected for either quantity or intensity, as the operator may desire. The rheophores are composed of large copper wire, heavily insulated with rubber.

I have made several improvements upon the original

Piffard battery of late, which, in my opinion, will increase the ease of working the instrument. They are not, as yet, fully perfected.

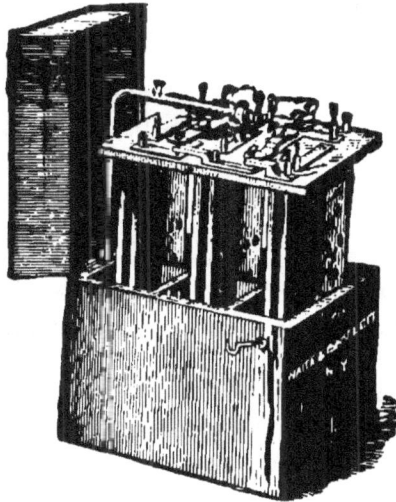

(a)

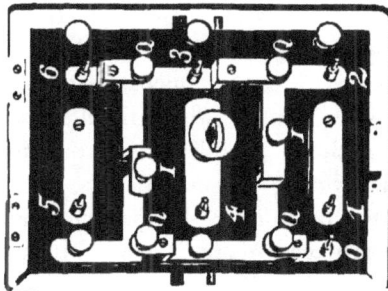

(b)

Fig. 40.—PIFFARD'S CAUTERY BATTERY. (a) The battery as suspended when not in action. (b) Arrangement of the top of this battery, showing the screws which regulate the connections between the different cells. The battery is rocked during its action to prevent extreme polarization. By making the movement slow or rapid, the heat of the loop or knife may be regulated at will.

It may be advisable to again impress upon you the fact that batteries designed for ordinary medical purposes are totally unfit for heating a cautery-loop or producing an electric light. A battery designed for cautery purposes is also totally unfit for other purposes in medicine.

In operations upon the tongue, nose, pharynx, uterus, vagina, rectum, and in some other regions, the galvano-cautery seems destined to supersede the scalpel and écraseur. No blood need be lost in amputations of considerable magnitude, provided the operation is skillfully performed. If the loop is employed, it is slipped when cold over the part to be removed. It can be adjusted, therefore, with every precaution against accident. After the current is turned on, the heat of the wire even can be regulated with great precision. Care should be exercised against drawing the wire too closely to the handle, and in selecting a wire which will not burn off or prove too large for the battery employed. As in all surgical procedures, this instrument should be handled by an expert. It is well for a novice to practice upon pieces of meat or bone until he familiarizes himself with the details of its use, in case he meditates performing an operation upon a human subject. When operations are to be performed within the mucous cavities of the body, the patient has frequently to be trained to tolerate the necessary manipulation. A dull red heat is preferable to a white heat in dividing vascular tissues, and it is very important that the division be slowly performed. When the skin is to be embraced within the loop, it is well to divide it first with a cautery-knife, and subsequently to adjust the wire.

The cautery-knife has been successfully used in removing cancerous growths within mucous cavities, in tubal pregnancy, in tracheotomy, in extirpation of the breast, and many other similar procedures.

An attachment to the cautery-battery, known as the "dome cautery," consists of a coil of platinum wire over a cone of porcelain. These may be of any size, and the porcelain cone may be omitted if deemed necessary. It may be employed in destroying hypertrophied tonsils, hæmorrhoids, polypi, nævi, epithelioma, etc.

The great advantage which the galvano-cautery has over the use of the knife is the absence of hæmorrhage and of great pain. The platinum knife can be made of any form desired. There is no limit to special forms of attachments which may be devised to simplify its use in different regions of the body.

In operating upon the tongue, needles may be passed through the organ in front of the site selected for the loop, so as to prevent slipping of the wire. Bryant, who has had an extensive experience in this operation, recommends a twisted wire rather than a large one. There is some reason to believe that the heat tends, moreover, to destroy (in the case of cancerous growths) the germs of the disease which might elude the knife.

GENERAL RULES GOVERNING ELECTRO-THERAPEUTICS.

Before we pass to the consideration of special methods of employing electricity in the treatment of disease, it seems to me advisable to suggest a few rules which may possibly aid you in deciding where and how to direct your treatment in any special case. There are, of course, some exceptions to each of these rules; but they are, nevertheless, sufficiently accurate to be used as guides in your practice:

1. Soak your electrodes in a weak solution of table-salt, not in simple water. This diminishes the resistance afforded by the skin at least fifty per cent.

2. Always press your electrode firmly and evenly against the part which it touches. This renders the cur-

rent employed an even one to the patient and assists in its conduction.

3. Put a milliampère-meter or a galvanometer, as well as the body of your patient, into circuit, and record all your observations, respecting the current-strength employed, from its scale. It is neither scientific nor accurate to simply record the number of cells employed. Cells grow weak by long-continued use, by polarization, and other causes. In case a faradaic instrument is employed, a galvanometer is useless; hence you should record the number either of centimetres or inches of the primary or secondary coil employed.

4. Always endeavor to apply one of the poles to the part which is diseased. The plates which I show you indicate the situation of the "motor points" of the head, trunk, and extremities. Such plates will enable you to direct your treatment to any special nerve or muscle.

5. Acquire, by frequent experimentation upon yourself, a knowledge of the effects of different current-strengths, the situation of most of the more important nerve-trunks, the formulæ of contraction of healthy nerve and muscle, and all other information necessary to the use of electricity in medicine.

6. Never use too strong a current upon a patient at the first sitting. It may frighten him, and he may never return. It is always best to begin with weak currents; in the majority of cases weak currents are indicated rather than strong ones.

7. If you have no galvanometer, the intensity of a galvanic current can be approximately determined by the burning sensation produced in the skin by the electrodes when they are applied to it.

8. The "polar method" is more painful when the faradaic current is employed than when the galvanic current is

used. It is not well to separate the poles of a faradaic machine too widely; pain is intensified, and no special benefit is gained by so doing. Remember that the faradaic current has no fixed polarity. A galvanometer will record the difference between the current produced by the "make" and "break" of the circuit only; hence it is of no value in determining the intensity of the faradaic current actually administered to a patient.

9. The "polar method" is absolutely requisite to electro-diagnosis when the galvanic reactions of nerve or muscle are being tested. It constitutes the best method also of administering the galvanic current for therapeutical purposes, because it is usually important that the anode or cathode exert its special influence upon the part diseased. The farther apart you place the poles, the less is the effect of the indifferent or neutral pole upon the part which you wish chiefly to influence.

Although clinical experience seems to prove that we obtain different results in the majority of cases by employing the anode or cathode upon the part to be influenced, I am inclined to question the correctness of the view that those effects are in any way dependent upon the direction of the transmitted current. We know that it is not possible to transmit an electric current in any one direction by means of animal tissues. Every current becomes diffused to a greater or less extent, as is illustrated in diagrams prepared by Erb and other authors upon electro-therapeutics. It is probably more correct to view the special effects obtained by employing the positive and negative poles of a galvanic battery as the effects of the poles themselves, rather than the result of the direction of the current.

10. Remember that the anode or positive pole of a galvanic battery is the sedative pole, and the cathode or negative pole is the stimulating or irritating pole. When the

cathode is made the indifferent pole, it is well to use a very large electrode.

11. Do not change the polarity of a current during its application to a patient any oftener than circumstances demand. As a rule, it is unnecessary to do so at all. It causes unnecessary irritation, which should always be avoided. In the treatment of neuralgia, diseased conditions of the brain or spinal cord, and painful points, it should never be done without some special reason. It is positively contra-indicated when catalytic effects are desired.

12. When galvanic currents to the head are indicated (especially if the current is to be sent through the brain), employ only those of moderate intensity (save in exceptional cases), and do not reverse the current unless there is good reason for so doing. When you read about thirty-cell currents being sent through the brain, it is safe to suppose that the battery was not of the most active kind, or that the ability of the patient to endure such a current was very exceptional. It is rare to meet with a patient who can tolerate a current of more than from three to six milliampères through the brain, and it is not safe to break currents of high intensity when employed about the head.

13. Static electricity will sometimes produce muscular contractions when faradaic currents will not. In hysterical conditions, some of the spasmodic diseases, sciatica, and organic spinal affections, it is well to try this form of electricity when galvanism fails to afford relief.

14. Respecting the duration of individual applications of electricity in its various forms, my experience teaches me that short sittings accomplish as much, and often more, than long ones. I seldom exceed five or six minutes, unless I am endeavoring to induce catalytic action, to benefit chronic articular rheumatism, etc.; or when I am employing general faradization, general galvanization, central gal-

vanization, electrolysis, the galvano-cautery, or other procedures which require a longer sitting. Frequently, thirty seconds to two minutes is all that is required when some particular part of the body is alone to be galvanized or faradized.

15. It is impossible to lay down any rule which will guide you in determining the frequency of the applications required by any individual case. It is seldom necessary to employ this agent oftener than every day, and three sittings a week will suffice in the majority of cases. If the disease is of a chronic type, it is often advisable to occasionally discontinue treatment for a few weeks, and then to renew it with vigor. Experience has taught me that the effects of electricity are more vigorous after such intermissions. It is often well to change from galvanic to faradaic, and again to static currents, from time to time, in the treatment of obstinate diseases which fail to progress satisfactorily.

16. I would advise you to be persistent in employing this agent when your judgment tells you that it is advisable to begin it. Many of the chronic forms of cerebral and spinal diseases are materially benefited and often completely cured by a proper course of electrical treatment which has been followed, with occasional intermissions, for some months during each year for several years.

17. As adjuncts to a course of electrical treatment, you will find massage, baths of various kinds, a change of climate, enforced rest in bed, and judicious internal medication, indicated in special cases. Delicate subjects, who suffer from neurasthenia, hysteria, persistent neuralgias, mental depression, sleeplessness, morbid fears, excessive " nervousness," rapid or extreme emaciation, profuse and persistent sweating of the palms or feet, dyspeptic symptoms, and the thousand other manifestations of debility,

are especially benefited by these adjuncts to a judicious use of electricity.

18. When simple excitation of motor or sensory nerves is demanded, the faradaic or static current is the best one to employ.

19. As a counter-irritant, and in the treatment of anæsthesia, dry faradization with a wire brush excels all other electrical applications, unless it be the use of static electricity.

20. In spasmodic diseases, in neuralgia, and other like conditions, galvanism and static electricity are alone of material service.

21. Interrupted galvanic currents are of service when muscular contractions of a forcible character are desired. When degeneration of a nerve exists, these can not be produced by the faradaic current.

22. The size of the electrodes modifies the density of the current directly. When large, the current is less dense because it is more diffused. The cathode should, as a rule, be larger than the anode when electrical applications are being made.

SPECIAL ELECTRO–THERAPEUTICS.

We have thus far discussed the various methods of employing electricity in a general way, and there remains for us to consider how we shall proceed in employing this agent when special organs are diseased. I would preface my remarks upon this field with the statement that the curative properties of electricity must, of necessity, be modified by the pathological conditions which exist in each individual case. The prognosis is naturally more grave in some conditions than in others.

For example, a patient who has motor paralysis which is due to *degenerative changes* in the cells of the anterior horns of the spinal gray matter will not usually recover the power of motion completely, while he may do so if the paralysis be due to a cerebral or spinal lesion which is not accompanied by degenerative nerve-changes. Again, all forms of functional nervous derangements are more amenable to electrical treatment (if judiciously employed) than are the graver results of organic disease of the nerve-centers. A muscle which has atrophied from disuse can usually be restored, while one which has wasted from imperfect nutrition (resulting from a degenerated nerve) may withstand all efforts to improve it. The therapeutical use of electricity is subject to the same influences as that of any other remedial agent, and the prognostic conditions are not always the same even among cases of the same nature.

In previous lectures I have given you many hints relating to the differential diagnosis which you will be called upon to make in nervous diseases, and enough has been said in reference to the anatomy and physiology of the nervous system to assist you in properly interpreting abnormal nervous phenomena. I shall therefore give you, in closing, directions only as to how to employ electric currents upon different parts of the body, without entering to any extent into the causation of the symptoms which you will be called upon to treat. Remember, however, that accuracy of diagnosis is the basis of cure in a large proportion of the cases which you will meet.

ELECTRICITY IN CEREBRAL AFFECTIONS.

Experiment has shewn beyond dispute that galvanic currents can be made to pass through the substance of the brain when inclosed within the skull. It is much less certain whether the same may be said of faradaic or static currents. The beneficial results which are obtained by the two latter (and possibly many of the effects of galvanism

as well) upon cerebral diseases are to be attributed, in my
opinion, chiefly to the alterations produced in the blood-
supply of the brain. Some of the most remarkable results
obtained by neurologists from the employment of electricity
upon the head itself or the cervical ganglia of the sympathetic are unquestionably due to an alteration produced in
the caliber of the cerebral vessels. I have never been convinced that *organic* lesions of the brain can be cured by the
direct use of this agent on that organ. On the other hand,
I am fully satisfied that the symptoms of cerebral hyperæmia and anæmia are directly influenced by galvanism and
static electricity. I believe that any unprejudiced mind
can be readily convinced of the scientific accuracy of this
conclusion. I have treated many patients (who gave undis-
putable evidences of basilar hyperæmia by the deflections
of the needle of a calorimeter), and have brought them to
a state of perfect health within a space of a few weeks by
galvanism of the head. The calorimeter confirmed the cure
in these cases by the absence of deflection which existed
before treatment was commenced. In some instances of
this condition static electricity proves a very valuable adjunct to galvanism. I will give you in detail a few of the
methods which, in my experience, may be employed in.
cerebral diseases with a prospect of great benefit to your
patient.

CEREBRAL HYPERÆMIA.—First ascertain by means of a
calorimeter the situation and extent of the congestion.
Test all parts of the head. When necessary, do so by separating the hair and bringing the poles as closely as possible in contact with the scalp. It is not necessary, as a
rule, to shave the head. In case very accurate observations
are demanded, this step may have to be taken—as, for example, when a cerebral tumor is suspected to exist.

At the nape of the neck, over the mastoid processes,

upon the temples, and over the forehead, no hair exists to interfere with the determination of the relative temperature of the two sides, or of different regions of the corresponding side. The calorimeter will aid you in diagnosis and treatment; if properly used, it is invaluable.

The following are the steps in treatment most generally useful:

(1) Apply the cathode to the nape of the neck, close to the skull, and the anode over the forehead. Make stabile applications for one or two minutes to each side of the forehead, the cathode remaining stabile. (2) Make labile anodal applications to the forehead transversely for one minute. (3) Move the cathode to the mastoid region of each side, place the anode centrally on the forehead, and continue each stabile application for from thirty seconds to one minute. This may make the patient dizzy. (4) Do not use a current which produces pain to the patient, but have as great intensity as he can comfortably bear. (5) Never reverse the current when the poles are on the head.

These applications may daily be alternated with "*insulation*" and the "*electric head bath*," if you possess a static machine. The sittings should occur daily until the symptoms are cured, and the calorimeter ceases to show its previous deflection.

It is sometimes well to *stimulate the superior cervical ganglion* by placing a small anode in the fossa behind the angle of the jaw, and the cathode on the seventh cervical spine, and to slowly interrupt the current. Caution must be exercised against employing too strong currents.

Finally, active *faradization of the limbs* is sometimes necessary, in order to draw the blood to the limbs. It is not well to employ this step if it causes an elevation of temperature.

The effects of this treatment should be to relieve the

pain or sense of fullness in the head, the vertigo on rising, the mental confusion or distress, the insomnia, and the

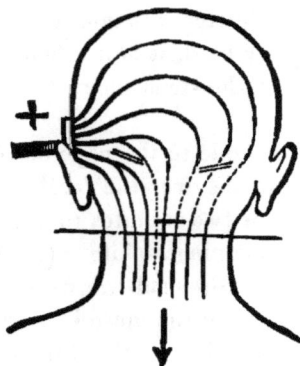

Fig. 41.—A Schematic Representation of the Distribution of an Electric Current applied unilaterally through the Head (after Erb). The anode (+) rests above the ear of the left side. The cathode (−) is supposed to be at the nape of the neck, and to exert its influence as far as the line drawn horizontally across the neck.

many other symptoms peculiar to this condition; and to steadily reduce the calorimeter deflections when the poles are in contact with homologous parts.

CEREBRAL ANÆMIA.—I should advise you to begin the use of very weak galvanic currents after an attack of embolism. I believe that currents of this kind sent transversely through the head from the temples, and occasionally in the antero-median plane, assist in absorbing the collateral œdema and cause a diminution of the collateral hyperæmia. I prefer to use the cathode on the side of the embolic obstruction when transverse currents are employed. In my opinion, it tends to promote absorption and to contract the vessels far more than the anode. The paralyzed muscles should be treated separately, by methods given in detail later.

Some four years ago Löwenfeld published some deductions relative to the action of galvanic currents upon the brain, based upon experimental researches. Although their accuracy has been justly called in question by authors of note, my own experience leads me to confirm them in part and to attach some importance to them. These conclusions were as follows: (1) Anode at forehead and cathode at neck causes contractions of the vessels of the pia; (2) anode at neck and cathode at forehead causes dilatation of the vessels of the pia; (3) when transverse currents are employed, the cathode causes contraction of adjacent vessels, and the anode dilatation.

When cerebral anæmia of a *general character* exists (as a manifestation of poverty of the blood, defective heart-power, etc.), general faradization, central galvanization, and static electricity by insulation are often of material benefit. The removal of the cause by judicious medication, etc., is, of course, vital to successful electrical treatment.

HEMIPLEGIA OF CEREBRAL ORIGIN.—A very large proportion of patients with hemiplegia from cerebral causes owe

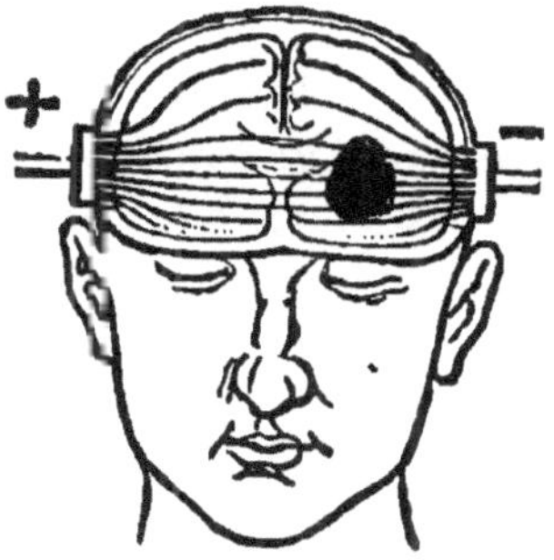

FIG. 42.—A SCHEMATIC REPRESENTATION OF THE COURSE OF ELECTRIC CURRENTS SENT TRANSVERSELY THROUGH THE HEAD (after Erb). The cathode (−) is represented as placed on the side of the lesion.

the paralysis of their limbs to hæmorrhage, softening, or embolism. The electrical treatment should be directed to both the brain and the muscles. It should not be commenced (save in the case of embolism) until a month has elapsed since the attack. Each patient's susceptibility to the agent should be carefully studied; and the strength of current employed should be modified accordingly. The muscles may be treated by faradization or galvanization, or by the static current (indirect sparks being drawn from the paralyzed limbs). The brain should be subjected to galvanization only, or to static insulation.

If the patient fails to show improvement within a month after the treatment has been daily applied, or if the improvement of the first few days is·rapidly lost in spite of continued treatment, the prognosis, as regards marked amelioration of the paralysis by electrical applications, is grave.

Hemianæsthesia is best treated by the wire-brush upon the dry skin in connection with the secondary faradaic current. I have also obtained some remarkable effects with the combined current (as before stated), and also with the static current, in cases where the faradaic current was ineffective.

Post-paralytic rigidity (occurring late) is the result, in most cases, of secondary changes within the spinal cord. The supervention also of pigmentation of the nails, œdema, a shiny skin, disease of the joints, and other evidences of trophic alterations, points to a serious and often permanent destruction of the nerve-centers.

Hints which have been given under the head of general electro-therapeutics will guide you in modifying the treatment according to the exigencies of each individual case. The remarkable improvement which some hemiplegics obtain through the instrumentality of electrical treatment should impress you with the necessity of employing it long enough to ascertain whether its continued use is indicated.

MONOPLEGIA OR MONOSPASM.—These conditions are

particularly indicative of cortical disease. The muscles affected are a guide to the convolution attacked. I have covered this field in previous lectures.* The indication in such a case is to improve, if possible, the nutrition of the diseased part directly by galvanism, and also to stimulate the muscles functionally associated with it. I employ for this purpose a " medium " electrode over the diseased convolution, the indifferent electrode being placed over the center of the sternum. It is my custom to employ both poles to the head for an interval of two minutes each at a sitting. The monoplegic limb may be treated by labile galvanic applications, the wire-brush and faradization, or the indirect spark by means of a static machine.

DUCHENNE's DISEASE.—The morbid changes in the nuclei of the medulla which accompany bulbar paralysis may, in some cases, be held in check for a while and the symptoms markedly improved by placing the positive electrode (of large size) at the nape of the neck and as close as possible to the foramen magnum, and applying the negative electrode (covered with absorbent cotton and attached to a long handle) successively to the pharynx, fauces, tongue, cheeks, and lips. As strong a current as the patient can easily endure should be used. The duration of the sitting should not exceed five minutes. It is well to complete the sitting by passing transverse currents through the neck, so as to excite the muscles concerned in deglutition. Some authors recommend the employment of currents through the head, both longitudinally and transversely.

ELECTRICITY IN SPINAL AFFECTIONS.

There are various ways of bringing the spinal cord under the influence of electrical currents. The method of

---

* See " Med. Record," May and June, 1834.

application selected in any individual case will depend
somewhat upon the symptoms which the patient presents,
and also upon the character and seat of the lesion. The
diagrammatic cuts of Erb, which illustrate the diffusion of
electrical currents, show in a graphic way the effects of
close approximation and wide separation of the poles. We
may also modify some of the morbid conditions of the
spinal cord by electrization of the extremities when the indifferent pole is placed over the spinous processes. It is
well to increase the size of the electrodes proportionately to
the strength of the current employed.

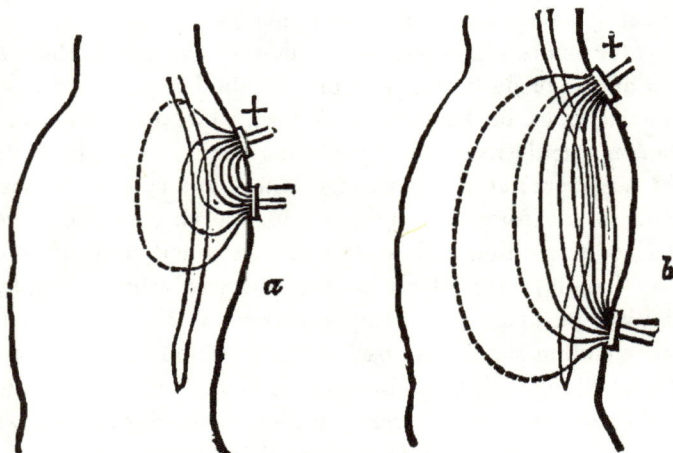

FIG. 43.—A SCHEMATIC REPRESENTATION OF THE DISTRIBUTION AND DEN-
SITY OF THE THREADS OF CURRENT WITH REGARD TO THEIR ENTRANCE
INTO THE SPINAL CORD (after Erb). In *a* the poles are placed near
each other. In *b* the poles are more widely separated. The size of the
electrodes shown in the cut is the same for both the anode and cathode.

Fig. 43 illustrates the effect of separation of the poles
when applications of electricity are made to the spinal column. Some of the threads of current depicted are ren

dered ineffective on account of their diffusion. This is made more apparent in Fig. 44.

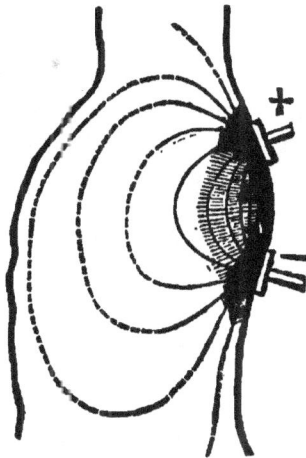

Fig. 44.—A Schematic Representation of the Density of the Current upon Application of the Electrodes .to the Same Surface and in Close Relation to Each Other (after Erb). The dotted lines indicate the ineffective threads of current. The shaded portion represents the zone of greatest intensity.

ELECTRIZATION OF THE SPINAL CORD.

To treat properly of the various methods which may be used when the application of electrical currents as a therapeutical measure for the relief of spinal diseases seems indicated, it would be necessary for me to enter into greater detail regarding spinal diseases than the time allotted to these lectures will permit of. I am reluctantly forced, therefore, to summarize somewhat hastily the main points which my experience with this agent leads me to indorse. Most of you are probably already familiar with the pathological changes which exist in connection with the more common diseases of the cord ; but, if any of you are not so, they should first be studied and thoroughly mastered before you can hope to successfully combat them.

Galvanic currents are of greater service in the treatment of spinal diseases than faradaic or static—chiefly on account of the depth of the tissues affected and the chemical and molecular changes which galvanic currents tend to induce.

Spinal electrodes should be of large size.

The applications may be either stabile or labile, the former being of the greatest benefit when the spinal lesion is circumscribed in extent, and the latter when a larger part of the spinal cord is affected. If labile applications are indicated, the movements of the electrodes should be made somewhat slowly.

In directing galvanic currents to the *cervical* and *upper dorsal segments* of the cord, it is well to place one electrode of medium size behind and below the ear alternately on the two sides of the neck, while the other is applied to the spine.

Points of tenderness to pressure along the spine should be subjected to stabile applications of the anode. They should be sought for in each individual case with care and separately galvanized.

The *strength* of the currents employed should be modified in individuals by the condition which is presented for treatment. Weak currents of from two to five milliampères act best, as a rule, when excessive irritability of the organ exists; chronic pathological conditions respond better to currents of greater intensity. I often use eight to twelve milliampères of current in chronic cases.

It is advantageous, in some subjects, to make *electrical applications to the limbs* when the cord is affected. Stimulation of the peripheral nerves and the muscles connected with the segments of the cord involved should be particularly aimed at, although the electrization of the skeletal muscles and the skin should not be exclusively confined to the limits thus indicated. It is my custom to employ the "*combined current*" (previously described) when applications to the limbs are thus made. This form of current is particularly indicated when the muscles exhibit a tendency toward atrophy. The electrode which rests upon cervical or lumbar enlargements of the spine should be of large size, while that used upon the limbs should be of medium size, so as to direct the combined currents to the nerves or muscles affected.

If *galvanism alone is used upon the limbs* in spinal disease, it is often beneficial to the patient to break the current by an interrupting electrode, or to reverse its direction by means of the commutator.

Some authorities advocate *faradization of the vertebral region and of the limbs* in conjunction with galvanic applications. I have seen in a few instances some remarkable effects follow the employment of the wire-brush alone in polio-myelitis of children, and I can see no reason to doubt its occasional efficacy in other forms of spinal disease.

In some unexplained way the excitation of muscular action and stimulation of the cutaneous nerves exert in many instances a remedial effect upon lesions of the spinal cord.

It is not always possible (as, for example, in polio-myelitis) to excite muscular action by faradism alone. In these cases interrupted galvanic currents, or the "combined current" (galvano-faradaic), will accomplish the desired end. I have repeatedly observed beneficial effects of this treatment in locomotor ataxia, and Rumpf has published some cases which sustain this view in which the wire-brush was used upon the arms and legs daily for about five minutes.

In all acute *inflammatory disorders of the cord* I deprecate the use of electrical applications. When the acute

6

stage has passed, or when the disease has assumed a chronic type, many of the effects of the disease (as, for example. muscular paralysis, rectal or vesical complications, incipient caries, anæsthesia, etc.) may often be greatly relieved by its judicious use. The current-strength employed in such cases usually varies from five to eight milliampères. The applications should be made daily. When possible, it is important that you localize early the seat of the structural lesion and concentrate the treatment, for a while at least, upon the segments of the cord involved. The muscles, skin, bladder, rectum, etc., should be separately subjected to the influence of electricity in case they exhibit a loss of function.

ELECTRICITY IN PARALYSIS OR PARESIS.

Hypokinesis may be due to many different conditions, hence its electrical treatment and prognosis must vary in accordance with the cause which excites it. You should remember that paralysis of a muscle is only symptomatic of other conditions, such as lead-poisoning, diphtheria, hysteria, mechanical pressure upon a motor nerve, severance of a motor nerve, destructive processes or inflammation within the motor cells of the brain or spinal cord, and changes in the vessels. All of these tend to impair either the generating power of a motor center, or the conducting power of a motor fiber.

Respecting the application of electricity to the seat of central lesions (i. e., lesions of the brain or spinal cord) in cases of motor paralysis, De Watteville pertinently remarks as follows:

"It is true that we have too often but a very imperfect idea of those processes in the nerve-centers upon which the symptom depends, and that we have no right to assume that the current has any specific curative influence upon any one of them; still, as a justification for central treatment in such cases, we may plead our very ignorance, we may urge the poverty of our therapeutical arsenal in arms wherewith to combat our enemy, and may also invoke the possibility of at least staying its progress by promoting nutrition of the surrounding portions of the nervous structures threatened by its invasion."

When the lesion directly affects the conductivity of a nerve, we have reason to believe that the direct influence of electrical currents upon the lesion tends to overcome the resistance offered to conduction by the disease-process, and facilitates the subsequent transmission of voluntary stimuli.

There are certain general rules that are applicable to the electrical treatment of paralysis of motility. These may be stated as follows:

1. The treatment should not be alone confined to the region of the paralyzed muscles.

2. The seat of the exciting lesion should be ascertained early, if possible, and subjected to the influence of this therapeutical agent in an intelligent way.

3. If the motor paralysis is accompanied by anæsthesia, hyperæsthesia, or other sensory disturbances, or if the vasomotor system of nerves be apparently implicated, the wire-brush may often be used with advantage upon the skin in the vicinity of the lesion, and also over the muscles paralyzed.

4. Faradaic currents (provided they excite muscular action), or the cathode-pole of a galvanic battery (with interruptions of the current), are of use in exciting the conductivity of the nerve-tracts affected. Static electricity is also of great utility in inducing muscular contractions, and is less painful than faradism or galvanism.

5. The "combined current" (galvano-faradaic) is chiefly of service in overcoming trophic disturbances, which often manifest themselves in connection with motor paralysis.

6. I prefer labile applications to stabile in applying either faradism or galvanism to the muscles. Stabile applications are preferable to labile when the brain, spinal cord, or peripheral nerve-trunks are to be influenced.

7. Never begin the use of electricity immediately after the onset of paralysis (when due to a central lesion). It is always best to wait until all danger of exciting a recurrence of the attack by stimulation of the nerve-centers has passed.

ELECTRICITY IN SPASMODIC AFFECTIONS.

HYPERKINESIS is frequently encountered as one of the varied forms of external manifestation of irritative and destructive lesions of the central nervous system. For example, it is by no means uncommon to observe convulsions (of the clonic or tonic type), tremor, muscular rigidity and contracture, etc., in connection with morbid changes in the brain and spinal cord. By these symptoms we are often assisted in determining the seat of the lesion, although, as De Watteville remarks, "the pathogeny of spasm is one of the most obscure problems in neurology." On the other hand—as, for example, in many instances of chorea, epilepsy, hysteria, etc.—spasm may exist without any apparent structural changes in the nervous system; being excited by some source of reflex irritation, such as phimosis, visual defect, uterine displacement, insufficiency of the ocular muscles, etc. In tetanus the exciting cause is generally found in one or more of the peripheral nerves. Sclerosis of the motor fibers of the lateral columns of the spinal cord is known to produce muscular contracture as a prominent symptom, probably because the inhibitory influence of the brain upon the reflex excitability of the spinal cord is arrested, or because the sclerosis directly excites the motor apparatus of the cord. The peculiar deformities produced by post-paralytic contracture and the eccentricities of gait and posture exhibited by patients suffering from tetanoid paraplegia (lateral spinal sclerosis) are illustrative of the diagnostic value of tonic muscular spasm in the course of some spinal affections.

Respecting the effects of electrical treatment of spasm, I am convinced that in some cases many methods must be tried without benefit before the right one is discovered. I have occasionally had brilliant results follow some particular method, and subsequently I have been utterly disappointed when it was tried upon some other patient with identical symptoms.

I think that in this class of subjects more depends upon your success in ascertaining and removing the cause than upon any electrical applications (valuable as they may be as adjuncts). The correction of an optical defect by glasses, the relief of ocular insufficiency by tenotomy or prisms, the operation of circumcision, the mechanical relief of a displaced womb, and many other such procedures, form the basis of an absolute cure in many cases which have been otherwise treated unsuccessfully. This fact is too often disregarded.

Electrical currents may be made to act upon these cases (1) as a sedative (chiefly the action of the anode and static insulation); (2) as a stimulant (the action of the cathode, the static spark, or faradism); (3) as a counter-irritant; (4) as a check to the progress of some peripheral or central morbid state (catalytic action); and (5) as an agent for the destruction of some neoplasm, induration, etc. (electrolytic action), or as a cautery.

I have lately come to regard static electricity (franklinism) as more generally applicable to spasmodic conditions (hysteria, torticollis, blepharospasm, tremor, contracture, etc.) than either faradism or galvanism. It seems, in my experience, to act more promptly, and to produce more last-

ing results, than the methods more commonly recommended
by authors.   I would advise those of you who decide to
purchase a static machine to try the effects of insulation,
the "electric wind," and the indirect spark (as the circumstances may indicate) faithfully before you resort to galvanization or faradization.   If good results are not obtained, you
can easily substitute for it the other forms of treatment at
a later date.   I should never regard any case as hopeless
until it had been thoroughly tried (after all reflex causes.
had been removed).   I have cured several severe cases of
tonic spasm of the muscles of the neck in a few sittings by
means of the indirect spark, and relieved many cases of suffering from other forms of spasm in a short time.

In EPILEPSY, the employment of galvanism alone has
never, to my knowledge, resulted in a complete cure, although some decided benefits have been reported from its
continued use.   There is, to my mind, a close relationship
in many cases *between epilepsy and ocular defect,* to which I
shall call your attention hereafter.   This element in the
causation of epilepsy certainly merits attention.   When all
defects in the visual apparatus have been corrected (in case
such exist), or when other reflex causes (such as phimosis,
for example) have been relieved, galvanism and static electricity may become valuable aids in controlling the subsequent convulsive attacks.   Latent hyperopia, astigmatism,
and insufficiency of any of the muscles of the eyeball may
(and, in my opinion, often do) excite epileptic seizures.   It
is absurd to expect of electricity, or any other therapeutical
agent, curative results when so important a source of irritation of the central nervous system is allowed to remain un-
corrected.

Rockwell's method of employing "central galvanization" in epilepsy does not, to my mind, equal in beneficial
effects the use of static insulation and the drawing of indi-

reet sparks from tbe neck and back of the patient. It is my custom, however, in some cases to employ both of these procedures, each being used alone during alternate weeks for a period of two or three months with daily sittings.

In CHOREA I have obtained the best results with static insulation and sparks.

My previous remarks respecting the relationship between defects in the organ of sight and epilepsy apply with equal force to this disease and all other types of functional nervous derangements. I shall discuss this subject more in detail later in the course when functional nervous diseases are being considered.

If galvanism is employed, it is best to subject the muscles affected with spasm to the action of the *anode*. The prognosis will depend somewhat upon the duration of the disease. The earlier you begin electrical treatment, the greater is the prospect of cure (provided all sources of reflex irritation have been successfully removed).

My experience with faradism in the treatment of chorea has been somewhat limited ; but the results obtained by me have not been so satisfactory as with static electricity.

In FACIAL SPASM (histrionic spasm) good results are occasionally obtained by following the plan of treatment suggested in connection with chorea; but treatment of the facial nerve alone is seldom satisfactory. I have one case at present under treatment, however, in which I have thus far had little, if any, success in my attempts to control the spasm. It is a case of long standing, and is therefore more rebellious to treatment than if it were not chronic. The patient has an ocular defect which it is difficult to correct perfectly.

In these cases I have obtained the best results by subjecting both the cortical centers for facial movements and the nerve itself to stabile applications of the anode (the cathode being placed on the breast-bone), and by treating the nerve at intervals with static sparks drawn from the affected portions of the face. The electrode for the head should be large. The duration of each daily sitting should not exceed five minutes.

NYSTAGMUS and BLEPHAROSPASM belong to the choreic type of diseases, and are best treated by electrical currents, provided they are seen before the condition has become chronic. The prospect of a radical cure steadily becomes less as time elapses. If static currents are employed, wooden tips to the electrodes should be used. I usually treat these cases as if the seventh nerve were involved in all of its branches. Sometimes it is well to place the anode upon the mastoid process and the cathode upon the closed eyelid. The current should be very weak at first; should be gradually increased until faint flashes of light are perceived; finally, it should be again decreased to the faintest perceptible point.

TORTICOLLIS, or WRY-NECK, when subjected early to static sparks or strong faradization, may often be cured very rapidly. Interrupted galvanic currents are also of material benefit in some cases.

The spinal accessory nerve is usually the one which is at fault. A rheumatic origin may often be detected. If so, judicious medication will tend to hasten the cure.

Some cases of wry-neck are associated with symptoms of paresis. These have, as you might suspect, a more serious prognosis. Electrical treatment will prove, as a rule, only palliative. Too often organic changes have already occurred in the spinal accessory nerve, the spinal cord, or the vertebræ. The duration of treatment should extend over a period of months.

SPASMODIC ASTHMA may often be benefited by galvanism of the neck. I have previously described the steps of this procedure. Its beneficial effects are probably due to changes induced in the vagi. Drawing of indirect sparks (by means of the static machine) from the anterior and posterior surface of the chest has proved, in my experience, an admirable preventive against such attacks.

Some patients have assured me that they experienced a sense of great comfort after each sitting, and that the frequency of the paroxysms of asthma has been perceptibly modified by them. My experience in the electrical treatment of these cases is as yet somewhat limited; but I am inclined to believe that greater benefit can be derived from it than from internal medication. Certainly it is worthy of a more extended trial, as an adjunct, if deemed wise, to other remedial measures, or as a substitute for them.

In TETANUS (both of the traumatic and idiopathic varieties) two cases of cure have been reported by Mendel, of Berlin. He employed galvanization and subjected the muscles affected with spasm to the stabile influence of the anode, the cathode resting over the spinous processes of the vertebræ. The applications were continued for fifteen minutes, and the currents employed were mild ones. Bartholow suggests, when speaking of these cases, that the effect of these applications was probably due "to the influence of the currents upon the sensory nerves, thus lessening the intensity of the reflexes." The cures were complete in about ten days.

Personally, I have not as yet been able to test the effects of the different forms of electrical currents upon a case of tetanus. To my mind, it would be very interesting, however, to observe the effect of static insulation and static sparks upon the spasms which occur paroxysmally in this disease. It is well known that this agent exerts a remarkable effect upon contracture of muscles. Thus far, to my knowledge, it has never been tried in tetanus.

SNEEZING, HICCOUGH, and COUGHING are spasmodic efforts of a reflex character. Occasionally they become distressing from their persistency. They may, in some instances, be relieved by faradization of the epigastrium, galvanization of the neck, and static electricity. De Watteville reports some curative effects of galvanization of the
nasal mucous membrane in chronic cases of persistent
sneezing.

ELECTRICITY IN DISORDERS AFFECTING SENSORY NERVE-TRACTS.

The discovery that different bundles of fibers which
help to compose the substance of the spinal cord serve to
convey sensory impulses only, and the later researches
which have also been made respecting the paths of conduction specially prepared for sensations of pain, touch, temperature, pressure, the muscular sense, visceral sensations,
etc., have a practical bearing upon both diagnosis and treatment.

Clinical observations go to show that, of the separate
and distinct types of sensation enumerated, some may be
partially or completely destroyed by diseased conditions
without impairing the others. Thus, for example, a patient
under certain conditions may be able to exercise his sense
of touch with normal acuteness and yet be rendered absolutely insensible to pain ; again, he may be unable to dis-
criminate between degrees of heat or cold (provided the
tests do not produce pain), although he retains unimpaired
sensory faculties in all other respects. We are, therefore,
forced to recognize a variety of types of anæsthesia as pre-
senting themselves for diagnosis and treatment.

The sensory functions may be either increased (*hyperæsthesia*) or diminished (*anæsthesia*).

Either of these states may be of *organic origin* (by which

we mean that structural changes in the nervous tissues accompany them), or of purely *functional origin*, in which case no structural changes can be shown to exist. Examples of the former are found in connection with central lesions (those of the brain or spinal cord), and with peripheral lesions of the sensory nerves or the organs of special sense, while examples of the latter are frequently encountered in connection with hysterical conditions, neurasthenia, cold, injury, imperfect capillary circulation, rheumatism, neuralgia, and many other morbid conditions.

In all forms of sensory disturbance the removal of the cause constitutes in many cases the basis of a cure, and the treatment will necessarily be modified by the causal indications.

Many suggestions which have previously been offered respecting electrical applications to the brain, spinal cord, and peripheral nerves are applicable alike to sensory as well as motor disorders when due to organic changes; hence, when this fact is borne in mind, it will be unnecessary to repeat what has already been given you.

In the treatment of ANÆSTHESIA nothing can surpass in its results the daily rise of the *wire brush* for about ten minutes over the regions affected. This form of electrode should be applied dry and with the *secondary coil* of a faradaic machine. The stabile electrode should be well moistened and pressed closely in contact with some distant point.

If *trophic disturbances* co-exist with anæsthesia, I have found the "combined current" (galvano-faradaic) to be more efficacious than faradism alone.

Static sparks and static insulation often act wonderfully in functional nervous diseases.

Static *insulation* has been previously described. It should be administered daily for from ten to thirty minutes.

If the "*indirect spark*" is employed (see Fig. 30), the length of the spark should be sufficient to be perceptible to the patient, and the duration of the application should seldom exceed five minutes. It is well to administer a fusillade of sparks to the region of the spine after each insulation, in case the sensory disturbances are dependent upon hysteria or neurasthenia.

I seldom employ the "*direct spark*" (Fig. 3) except in the treatment of organic disturbances of sensation or motion.* This form of administration should be used with extreme caution if the generating machine is a powerful one.

The "umbrella" electrode furnishes an agreeable and effective method of concentrating static electricity to the head of the patient. The sensation is one which resembles that of a strong breeze circulating through the hair.

HEMIANÆSTHESIA (whether of cerebral or spinal origin) is often benefited by cutaneous faradization of limited portions of the area affected—a point first observed by Vulpian, who employed this method with marked success.

TROPHIC DISORDERS may occasionally manifest themselves, often in the skin, nails, hair, and muscles, when sensation is markedly affected. One such case (suffering from locomotor ataxia) was lately placed under my care. The fin-

---

* Since these lectures were delivered I have at last perfected an improvement upon the Holtz static machine, upon which I have been working for some time, and I have had one of this pattern built for my own use by Waite & Bartlett, of New York. I intend soon to present this improvement formally to the profession with appropriate cuts and a description of the various modifications made. My own machine after this model gave me a ten-inch spark during a "muggy" day in July, when most static machines would produce but a feeble spark. It is the most effective instrument of the kind which I have yet seen. It consists of nine twenty-four-inch glass plates, six of which revolve. All the joints of the case are hermetically sealed with soft rubber.

gers of both sides were almost destitute of sensibility to pain, and tactile sensation was impaired. The nails were thickened, loosened for half of their length, and deeply pigmented (as if stained with iodine). The terminal phalanges were " clubbed," the nails being bent almost into a semicircle. The skin was thickened and very hard under the loosened nails. The " combined current" (galvanofaradaic) with a wire-brush electrode caused decided improvement within a few weeks.

NEURALGIA (when of idiopathic origin) is more successfully treated to-day by electricity than by any medicinal agent. In many instances it is cured in a few sittings.

It is well to bear in mind, however, the fact that neuralgic pains are very often symptomatic of causes more or less remote from the affected nerve, and that a permanent cure is impossible in many instances as long as that cause actively exists. Defective teeth, morbid processes in the bones, pressure upon a nerve, organic changes in the nerve itself, toxic diatheses, rheumatism, gout, reflex irritation from the eye, uterus, digestive tract, ovaries, etc., cardiac and pulmonary disorders, and many other morbid conditions, may be enumerated as among the exciting causes of neuralgia.

Respecting the electrical treatment of neuralgic pains (*per se*) the following deductions may prove of some advantage to you :

1. If points of tenderness to pressure (*puncta dolorosa*) exist along the course of the affected nerve or its branches, it is well to subject them to stabile galvanic applications of the anode, the cathode being placed at a neutral point.

2. The anode should be made to cover as large an area as possible.

3. The duration of the sitting should not exceed five minutes, save in exceptional cases. The sittings may be repeated several times a day if necessary.

4. As a rule, it is unwise to break the current. In obstinate cases the current may occasionally be reversed, without changing the poles, by means of the commutator.

5. Faradization of the nerve and the use of the wire brush upon the skin have been recommended when galvanism proves unsuccessful in arresting the pain. It should not be used (in my opinion) until galvanism has been thoroughly applied.

6. It is well in obstinate cases to direct the applications of galvanism to the central origin of the affected nerve, as well as to its peripheral distribution.

7. Static electricity often produces marvelous results in neuralgia. I have more faith in it as a cure for sciatica than in any other remedial agent. It should be applied (by the "spark" method) over the affected nerve. One sitting has, in my experience, frequently arrested severe pain. It gives immediate relief, in most cases, to muscular rheumatism also, and to lumbago. Sufferers from muscular and neuralgic pains are perhaps as frequently encountered by the physician as any class, and static electricity should highly recommend itself to your confidence for such cases. The expenses of the outfit, and the fact that the machine is too large for transportation, will probably prevent its general use by the profession; but, until its effects upon a patient have been tried, I would caution you against expressing an unfavorable opinion, even if galvanism, faradism, and medicinal treatment have proved powerless to relieve the suffering.

8. The operation of electro-puncture of a nerve for the relief of neuralgia has proved of benefit in the hands of some neurologists; but it is an operation which, if injudiciously employed, will produce electrolysis, and serious results may follow its use.

9. The electrical treatment of various other forms of

pain is similar to that of neuralgia. The judgment of the physician should be exercised regarding the position and size of the electrodes, the variety, strength, and duration of the current employed, and various other minor points suggested by the condition of the subject.

10. *Visceral neuralgias* (as, for example, the conditions known as hemicrania, migraine, gastralgia, enteralgia, hepatalgia, etc.) are often relieved by electricity, irrespective of the reflex or constitutional condition which induces the morbid state. The removal of the exciting cause, however, will greatly assist in making the cure a radical one. I have long taught in my lectures that I had yet to encounter a patient who had suffered for years from migraine who had not some defect in the eye or its muscles as an exciting cause. Experience leads me still to strongly assert this as my conviction. The same cause is very frequently manifested by paroxysms of spinal pain—peculiarly so at two points, viz., between the scapulæ, and at the junction of the last lumbar vertebræ with the sacrum.

The currents which act best upon these cases are the galvanic and static. I have in two instances employed faradism in gastralgia with good results, but I regard it as inferior to galvanism or franklinism.

In treating the abdominal viscera by galvanic currents, one rheophore may often be attached with advantage to a rectal electrode, and the other to a large electrode placed over the organ to be influenced. I do not believe that polar effects are to be particularly aimed at. In some cases, an occasional substitution of the "combined current" (galvano-faradaic) for galvanism makes the improvement of the patient more rapid.

Static applications to the abdomen are best made by employing indirect sparks of about two inches in length. Long sparks are not borne well by sensitive subjects. If patients are subjected to static insulation alone for twenty minutes daily, or to the electric spray over the abdomen, relief is often afforded and the application is painless. The clothing need not be removed in making applications of franklinism by either of these methods — a point which renders the treatment particularlyagreeable to ladies.

ELECTRICITY IN DISEASES OF THE CERVICAL SYMPATHETIC, THE VASO-MOTOR SYSTEM, AND ALLIED NEUROSES.

THE CERVICAL SYMPATHETIC is undoubtedly, in rare cases, the seat of isolated morbid changes; but, as Erb remarks, these cases "constitute pathological curiosities." The morbid conditions which have been detected embrace inflammation, compression, traumatism, rheumatic conditions, etc. Such conditions may create either irritation of the sympathetic system or paralysis of its functions, or both simultaneously in different parts of the body.

Irritation of the cervical sympathetic produces pallor of the face and neck upon the affected side, with a sense of coldness in the parts. The pupils are dilated, the temporal arteries exhibit increased tension, the power of accommodation and the reaction of the pupil to light are both impaired, the eyeballs protrude slightly, and the secretion of sweat is diminished.

Paralysis of the cervical sympathetic induces the opposite conditions. The skin is red and hot, the patient suffers from a sense of heat in the skin, the pupils are contracted and exhibit normal reactions to light and accommodation of vision; the eyeball does not protrude, there are often headache and vertigo, the secretion of tears and sweat is increased, and the pulsation of the carotids is excessive.

In the electrical treatment of these opposed conditions Erb recommends stabile applications of the anode (with a strong current) until a change in the pupil is observed, if the condition of irritation exists. The same author suggests the use of the cathode with a feeble current, frequent interruptions, and occasional reversal of the poles, if the paralytic state is present. He places the "indifferent" electrode upon the spine. He also suggests applications of the wire-brush, or labile galvanic currents, to the skin of the face and neck.

To the views of this author I would urge the advantage of trying the effects of static insulation and sparks directed to the neck and face.

ANGIONEUROSES OF THE SKIN may assume one of two forms, viz., *spasm* or *paralysis*. They are most frequently observed in connection with neurasthenia and in hysterical patients. The abnormal contraction or relaxation of the vessels may cause (1) modifications in the color and the general "feel" and sensibility of the skin; (2) subjective sensations of heat, tingling, formication, etc.; (3) disturbances of perspiration; (4) awkwardness of movement of the part (especially in the hands); and (5) many reflex symptoms referable to the viscera.

Unnatural conditions of the vessels of the skin (spasm or paralysis) are most frequently observed in the upper limb, less frequently in the lower limb, and least often in the face and neck. They may be excited by a variety of causes—such as fatigue, excitement, menstrual disturbances, malaria, exposure to cold, the effects of poisons, and direct irritation of the skin itself.

I have seen the skin (especially of the fingers) made as white as chalk in some cases, and in others rendered cyanotic, by *spasm of the vessels*. The muscles of the papillæ of the skin may participate in the spasm and produce the so-called "goose-flesh" appearance. Pain, tingling, formication, partial anæsthesia, and other disturbances of the sensory apparatus may occur as sequelæ to the vascular spasm.

Paralysis of the cutaneous vessels leads to directly oppo-site conditions. The skin may be made intermittently or permanently red, and feel unnaturally hot and extremely sensitive. Subjects so afflicted frequently suffer from in-somnia, headache, disturbed heart-action, excessive perspi-ration, vertigo, and other visceral manifestations of irrita-bility.

Respecting the electrical treatment of angiospasm and angioparalysis, the general rule may be given that weak or moderate applications of faradism or galvanism to the affected part act best upon dilated vessels, and stronger currents upon those affected with spasm.

Applications of static electricity are often very beneficial to neurasthenic and hysterical subjects. Personally, I be-lieve this method of treatment surpasses any other in its effects upon this class, although it is well to alternate with galvanism and faradism when a case proves obstinate to treatment.

When any of the methods suggested are employed, it is well to subject both the vaso-motor centers and the nerve-trunks which supply the affected regions (as well as the parts directly) to the influence of electrical currents.

NEURASTHENIA.

By this term we mean the condition of *nervous exhaus-tion*. It may be manifested in a variety of ways. Its symptoms will depend upon the type which exists—cere-bral exhaustion or spinal exhaustion—and also upon special idiosyncrasies peculiar to the patient. No two cases exhibit identical manifestations of nervous depression. Some pa-tients who are suffering from cerebral neurasthenia mani-fest its effects in the voice, others in mental disturbances. The heart's action may be alone disturbed in some cases, the stomach may give out in others, some may complain alone of muscular troubles, some may notice its effects in the eyes, some are rendered sleepless, a few complain alone of skin disturbances, and so on throughout the different parts of the entire human organism.

You can understand how these apparently discordant facts may be reconciled when you consider the fact that, by means of the brain and spinal marrow and the nerves which unite these centers to the different parts of the body, we are enabled to see, hear, taste, smell, appreciate touch, swallow, breathe, and perform voluntary muscular acts. It is by means of our nerves alone that the heart beats; the digestive processes go on, without our knowledge or control, through the same agencies; the blood-vessels contract and dilate in accordance with the demands for blood telegraphed to the nerve-center by different organs and tissues; and every process pertaining to life is thus automatically regulated. Now it is easy to see at once how a debility of so complicated an electric mechanism as the nerve-fibers and the nerve-cells of a living animal are, can upset all or any one of the individual functions enumerated. Many of our houses are furnished to-day with electric bells by means of wires distributed in the walls. In some houses we light the gas-jets, and even the rooms themselves, by means of the same subtle fluid. When the battery becomes weak, or when the wires are disarranged or broken, what may be the results? Some of the bells may cease to ring when the button is touched, while others work properly. Perhaps the electric light may fail in some rooms and burn with its accustomed brilliancy in others. The gas-jets may not be properly ignited. So it is with the nervous apparatus of man. From the same cause one patient may have nervous dyspepsia, another sleeplessness, a third sexual debility, a fourth weakness of the eye-muscles, a fifth disturbances of the skin. It is needless to multiply illustrations.

CEREBRAL NEURASTHENIA (*brain exhaustion*) may be indicated by one or more of the following symptoms: Tenderness of the scalp; pains in the head; fleeting neuralgias; sleeplessness; vertigo; a tenderness and pallor of the gums; abnormal sensitiveness of the teeth; blanching of the hair; flushings of the face; dilatation of the pupils; idiosyncrasies in regard to food and external irritation; mental depression and melancholia; defects in memory; a morbid craving for alcohol; a decrease in intellectual capacity; a buzzing or ringing in the ears; specks before the vision; abnormal and imaginary impressions of taste or smell; morbid fears of various kinds; sick headache; dryness of the skin and the mucous surfaces; weakness of the muscles; numbness in the limbs; thickness of speech; and mental excitability, irascibility, or loss of emotional control.

These symptoms, in many cases, are but the manifestations of weakness. The electric batteries of the brain (those minute organs known as the "brain-cells") are feeble or uncertain in their action. They are incapable of performing the offices for which they were created. They are not diseased (in a medical sense), but they are weak and liable to become so sooner or later. I have known sufferers of this type to be precipitated into a condition of incurability by mental alarm, excited, in some instances, by an opinion of an unfavorable kind made by physicians respecting a prospect of recovery. Again, it is well known that insanity may arise as a consequence of the loss of sleep often experienced by these subjects, and by " brooding over their symptoms," whose significance they fail to properly understand. I recall several cases where a patient was with difficulty convinced that some special type of malady was not about to attack him, because in reading a medical work his attention had been called to the significance of some special symptom which he was sure he had personally experienced. If medical students, who possess vivid imaginations, can become (as they often do) victims to imaginary diseases whose symptoms they have been studying, is it to be wondered at that the weak and highly organized sufferers from neurasthenia are especially prone to become impressed by this form of delusion ?

SPINAL NEURASTHENIA (*spinal exhaustion*) signifies an exhausted state of the cells which help to form the spinal cord. The cord itself is of about the size of an ordinary lead-pencil, and is sixteen inches in length (much shorter than the backbone). It is composed of millions of nerve-cells and distinct bundles of nerves. Some of these nerves pass through it to reach the brain above, while others become united to the spinal cells and pass no farther. The cells of both the brain and spinal cord are practically electric batteries; and the nerve-fibers are the wires by which they are connected with the different organs of the body, the muscles, skin, joints, and viscera. This wonderfully constructed organ is under the control of the brain, but is capable of exerting, under certain circumstances, a control over all acts which are termed " reflex acts," because they are, to a greater or less extent, independent of the will. It serves also as a means of conduction to the brain of our perceptions of pain, temperature, and touch, and of motor impulses sent out from the brain-cells to the muscles of the limbs and body.

Now, when the cells of the spinal cord become exhausted, the symptoms produced differ markedly from those already mentioned as indicative of brain-exhaustion. Among its chief manifestations may be mentioned the following : A general tenderness of the skin to touch or pressure ; tenderness along the spine or over certain limited portions of the spine ; irritability of the breasts, ovaries, and the womb in females; fleeting pains of a neuralgic type in various parts of the body; an extremely rapid or slow pulse, which fluctuates widely during periods of excitement or fatigue; attacks of palpitation of the heart; dryness of the skin, or in many cases the reverse, excessive perspiration of the hands and feet; sudden startings on going to sleep; muscular twitchings in one muscle or a group of muscles; chilliness and creeping sensations along the spine; numbness or abnormal sensations of heat in the skin of the body or limbs; itching of the skin; eruptions upon the skin, chiefly of the type of eczema; frequent gaping, yawning, or stretching; frequent seminal emissions; weakness of the bladder and rectum; and disturbances of the digestive functions.

The distinction between cerebral and spinal neurasthenia which has been shown to exist can not be made in each and every case, because various combinations of the symptoms of the two are often encountered in the same individual. A prominent author upon this type of diseases very aptly compares the nervous system of man to certain mountainous regions, since it causes so many echoes and reverberations. He says: "An irritation at one point may be transferred to any other point, following the paths of least resistance and making itself felt in those parts that are least able to resist molecular disturbances. Thus, for example, seminal emissions and spermatorrhœa, when they arise through abuse or through spinal cord disease, almost uniformly react on the brain—robbing the sufferer of courage and manliness, exciting various phases of morbid fear, of which I shall speak, with aversion of the eyes and countenance."

I have known a decayed tooth to cause persistent earache, and in one case to cause the corresponding eyebrow to become white. In male children a tight foreskin not infrequently creates sufficient irritation of the sexual organs to induce spasms or paralysis of the lower limbs by an indirect effect upon the spinal cord. I have cured some patients, who have come to me for relief from persistent and excruciating attacks of neuralgia, by a correction of some defect in their eyes by means of glasses. The strain and irritation produced upon the brain by the involuntary efforts made by these patients to see objects with distinctness or to read and write had reacted upon the nervous system; and it would have continued so to react till death if the cause of the irritation had not been removed.

It may be well to consider a few of the more prominent manifestations of nervous exhaustion separately. Among these, sleeplessness, a defect in vision known as asthenopia, sexual weakness, headache, an unnatural dryness of the skin and mucous surfaces or profuse sweating of the hands, and morbid fears or melancholia, deserve special mention.

INSOMNIA.—Sleeplessness may assume different forms. Some of those afflicted have difficulty in getting asleep; some awake after a few hours of slumber and remain so until daylight; a few find themselves overpowered with a desire for sleep during their working hours, when their business will not admit of it, and at night can not obtain sleep except under narcotics. I have had patients who have told me that they spent most of their nights for years in writing to friends, riding in the horse-cars, or walking the streets for amusement because they could not sleep. It is safe to assert that persistent insomnia, extending over a period of weeks or months, indicates disease of some kind. In patients who have passed the age of fifty, or in younger persons who have indulged to excess in alcohol, it is often due to a type of kidney disease, to detect which repeated examinations of the urine are required. This form of trouble is known as the "granular" or "contracted kidney"; and insomnia, frequently combined with headache,

is one of its most prominent symptoms. Obstinate sleeplessness is the cause of many a suicide, too often the start-
ing-point of the opium and chloral habit, and surely the
destroyer. I would caution you against allowing this symptom to remain uncontrolled for any length of time, and to
avoid the use of all forms of narcotics as long as possible.
The chains of intemperance are but silken threads when
compared to those of the opium or chloral habit.

ASTHENOPIA.—This type of defective vision can not be
relieved by ordinary glasses, nor does it respond quickly to
the customary suggestions of gymnastics, horseback-riding,
etc. It is due to a peculiar weakness of the muscles which
control the movements of the eyeballs, and it manifests itself chiefly as a sense of extreme weariness when the eyes
are steadily employed for short periods of time. It is an
indication of neurasthenia, and is of great diagnostic value.
In severe cases it becomes necessary to divide the tendons
of the stronger muscles of the eye, in order to relieve the
weaker ones of a strain. It is common among near-sighted
and far-sighted persons. I have seen patients who could
not sew for five minutes at a time from this cause, and
others who would be made sick by attending a theatre, pic-
ture gallery, or other places of amusement.

HEADACHE.—Many attacks of this character are undoubtedly to be attributed to imprudences in eating, exposure, or fatigue. But I believe that many of those who
are periodically afflicted in this way owe their suffering to
a lack of tone in the muscular coat of the blood-vessels of
the brain, consequent upon some of the causes of neurasthenia mentioned. I have seen a large number of instances
where the *eyes were the cause of such headaches*, and where
the adaptation of glasses has brought immediate relief.
The medical profession are rapidly becoming enlightened
upon this fruitful cause of pain. It is well also to examine

the urine when persistent or periodical headache occurs, as it may be a symptom of kidney disease. Some neurologists believe that the so-called "sick headaches" are to be regarded as but a modified form of that condition which produces epilepsy. A volume might be written upon this symptom alone.

DRYNESS OR UNNATURAL MOISTURE OF THE SKIN.— Some nervous patients suffer from an unnatural dryness of the skin, the throat, and the nose. They are also liable to experience dyspeptic symptoms at the same time, which is probably due to similar changes in the lining of the stomach. This dryness may be accompanied also by an itching of the affected parts or an attack of eczema. A burning sensation is sometimes produced. I was once consulted by a patient who had for years been in the habit of incasing himself in flannel and putting on flannel stockings before he retired, in order to overcome a sense of burning in the skin which followed the contact of cotton or linen with any part of his body. I recall a case where the feet were once frost-bitten, and the patient has never since been able to walk upon a carpeted floor on account of a burning sensation which immediately follows. He takes off his shoes as the last step before retiring.

On the other hand, many patients afflicted with neurasthenia suffer from a profuse sweating of the palms of the hands. This is accompanied in some instances by a flushing and redness of the face, neck, and ears. The nails may become unnaturally soft or brittle.

MORBID FEARS.—This peculiar manifestation of nervous exhaustion may assume one of several types. Attempts at classification of these morbid fears have been made by some authors, such as fear of lightning, of places, man and society, solitude, accident, etc., and special names have been applied by them to each of these types. Fears of this kind

7

may be present without any other manifestation of mental impairment. They are usually uncontrollable, in spite of the fact that the patient may exhibit a knowledge that they are groundless and absurd. They seem to take full possession of a being, and to cause mental torture of an extreme kind.

Finally, melancholia is not an infrequent symptom of neurasthenia. It may be accompanied by paroxysms of laughing, weeping, and hysterical phenomena.

Now, in the treatment of neurasthenia, electricity is one of our most effective agents. After the exciting cause has been discovered and the possibility of its continuance removed, we may safely begin the use of electricity with the brightest prospect of a radical cure. General faradization, central galvanization, and the use of franklinism are particularly of service. Of the-latter I can speak in the highest terms. Neurasthenic patients often feel its beneficial effects immediately. It should be applied daily by the insulation method, the electric wind, or the static spark, as the circumstances of the case seem to indicate.

My remarks made in a previous lecture respecting massage and other adjuncts to electrical treatment are particularly applicable to this class of patients. No effort should be spared in your treatment to promote constitutional vigor by exercise, judicious feeding, good hygienic surroundings, and the like.

You, as physicians, will have to decide such matters for yourselves, and advise your patients respecting them with references to the symptoms which are to be combated. Do not trust too implicitly in electricity alone (valuable as it may be as a means of cure). Active employment will be necessary to a cure in some cases; absolute physical and mental rest will aid in others; some will require travel or a change of surroundings; the organs of the body will often demand special attention with a view of properly regulating their functions; and many other similar problems will have to be decided before a cure can be predicted by the aid of electrical agencies.

In bringing this course of lectures to a close, gentlemen, I can not but feel that much has been, of necessity, omitted by me which would be of benefit to you. A more complete course upon electro-physics, for example, would have been given had I not felt that text-books on physics would furnish you with the requisite knowledge when needed, and that the time allotted by me to the consideration of electricity could be better spent in dealing with the practical applications of this agent in the diagnosis and treatment of disease. In these respects even this course is far from complete. It is but a hasty sketch of the more important facts.

It is well for each of you to bear constantly in mind that electricity as a therapeutical agent is yet in its infancy. Facts are being daily brought to light, however, which will aid us in employing it upon the sick to better advantage, and in obtaining more uniform results. As fast as new discoveries are published they will naturally be subjected to tests by those laboring in this field, and, if their value is proved, they will in time become generally recognized and employed by the profession.

Perhaps some of this class of students may be among the number who are destined to promote the growth of this department of therapeutics by their inventive faculty, original research, or clinical observation. I trust that it may prove so.

If I have succeeded in awakening your interest sufficiently to induce any of you to pursue this line of study further with intelligence, and thus to give possible benefit to others, I shall feel amply repaid for the many hours of thought and manual labor that I have personally spent in trying to modify and improve existing electrical appliances.

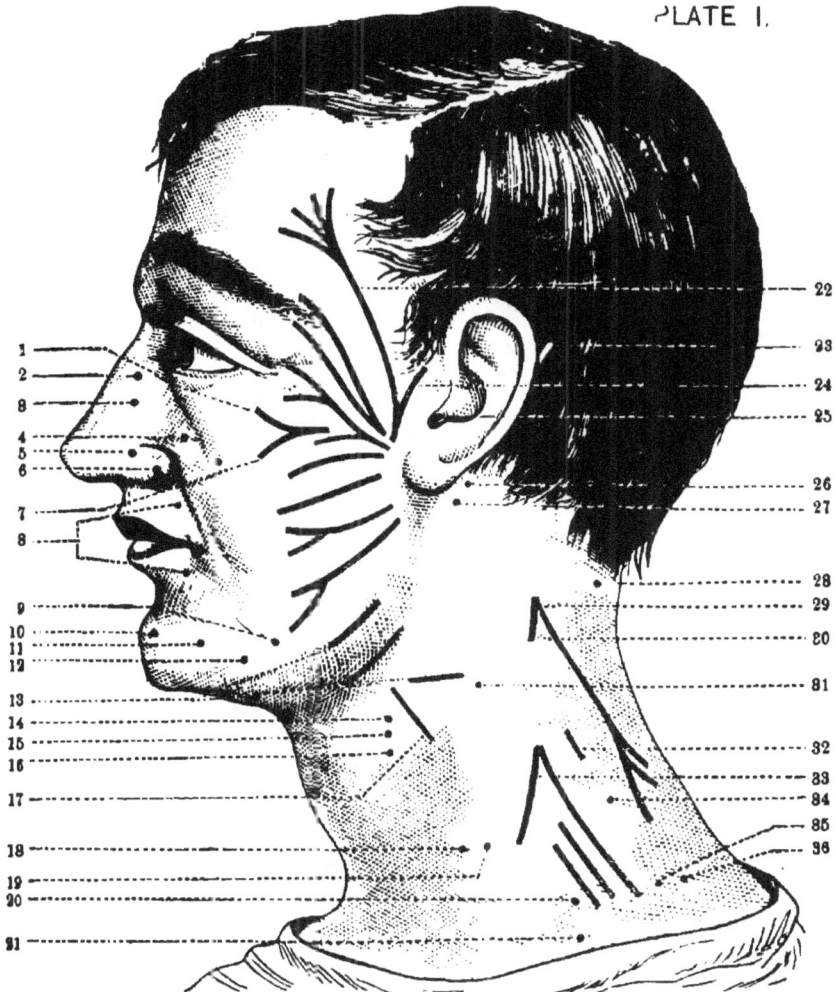

PLATE I.

A DIAGRAM OF THE MOTOR POINTS OF THE FACE, SHOWING THE POSITION OF THE ELECTRODES DURING ELECTRIZATION OF SPECIAL MUSCLES AND NERVES. THE ANODE IS SUPPOSED TO BE PLACED IN THE MASTOID FOSSA, AND THE CATHODE UPON THE PART INDICATED IN THE DIAGRAM.

1, m. orbicularis palpebrarum; 2, m. pyramidalis nasi; 3. m. lev. lab. sup. et nasi; 4, m. lev. lab. sup. propr.; 5, 6, m. dilator naris; 7, m. zygomatic major; 8, m. orbicularis oris; 9, n. branch for levator menti; 10, m. levator menti; 11, m. quadratus menti; 12, m. triangularis menti; 13, nerves—subcutaneous, of neck; 14, m. sterno-hyoid; 15, m. omo-hyoid; 16, m. sterno-thyroid; 17, n. branch for platysma; 18, m. sterno-hyoid; 19, m. omo-hyoid; 20, 21, nerves to pectoral muscles; 22, m. occipito-frontalis (ant. belly); 23, m. occipito-frontalis (post. belly); 24, m. retrahens and attollens aurem; 25, nerve—facial; 26, m. stylo-hyoid; 27, m. digastric; 28, m. splenius capitis; 29, nerve—external branch of spinal accessory; 30, m. sterno-mastoid; 31, m. sterno-mastoid; 32, m. levator anguli scapulæ; 33, nerve—phrenic; 34, nerve—posterior thoracic; 35, m. serratus magnus; 36, nerves of the axillary space. In this text m. = muscle; n. = nerve.

PLATE II.

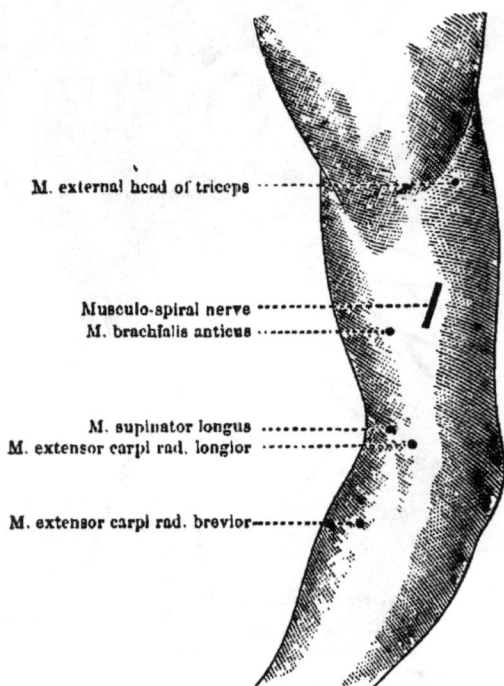

M. external head of triceps ----------

Musculo-spiral nerve ----------
M. brachialis anticus ----------

M. supinator longus ----------
M. extensor carpi rad. longior ----------

M. extensor carpi rad. brevior----------

THE MOTOR POINTS ON THE OUTER ASPECT OF THE ARM.

PLATE III.

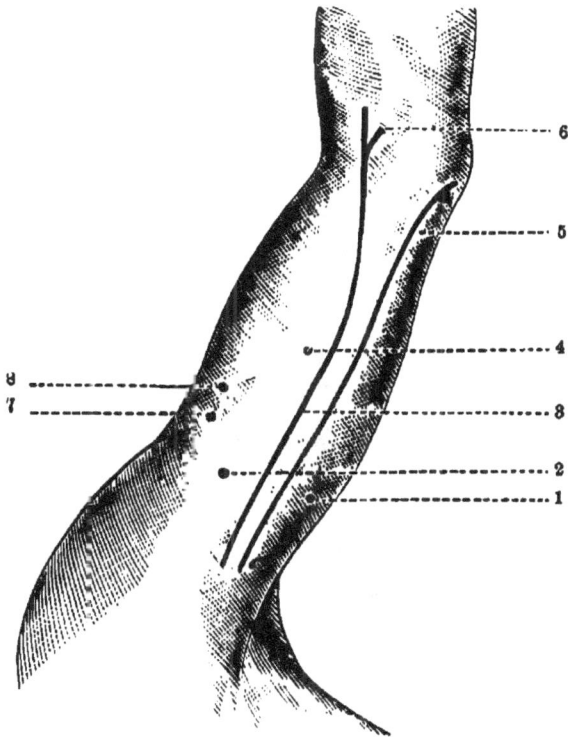

THE MOTOR POINTS ON THE INNER SIDE OF THE ARM.

1, m. internal head of triceps: 2, musculo-cutancous nerve; 3. median nerve; 4, m. coraco-brachialis; 5, ulnar nerve; 6, branch of median nerve for pronator radii teres; 7, musculo-cutaneous nerve; 8, m. biceps flexor cubiti.

PLATE IV.

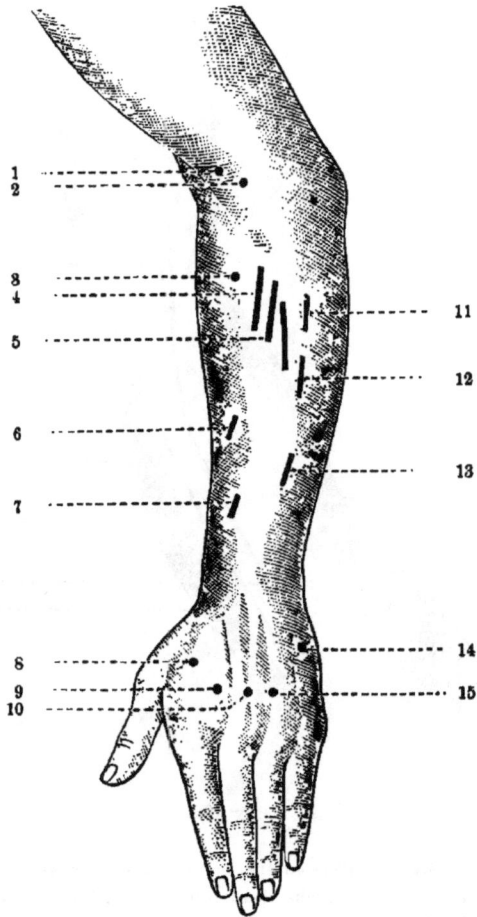

THE MOTOR POINTS ON THE EXTENSOR (POSTERIOR) ASPECT OF THE FOREARM.

1, m. supinator longus; 2, m. extensor carpi rad. longior; 3, m. extensor carpi rad.
brevior; 4, 5, m. extensor communis digitorum; 6, m. extensor ossis. met.
pol.; 7, m. extensor primi. internod. pol.; 8, m. first dorsal interosseous; 9, m.
second dorsal interosseous; 10, m. third dorsal interosseous; 11. m. extensor
carpi ulnaris; 12, m. extensor min. digiti; 13. m. extensor secund. internod.
pol.; 14, m. abduct. min. digiti; 15, m. fourth dorsal interosseous.

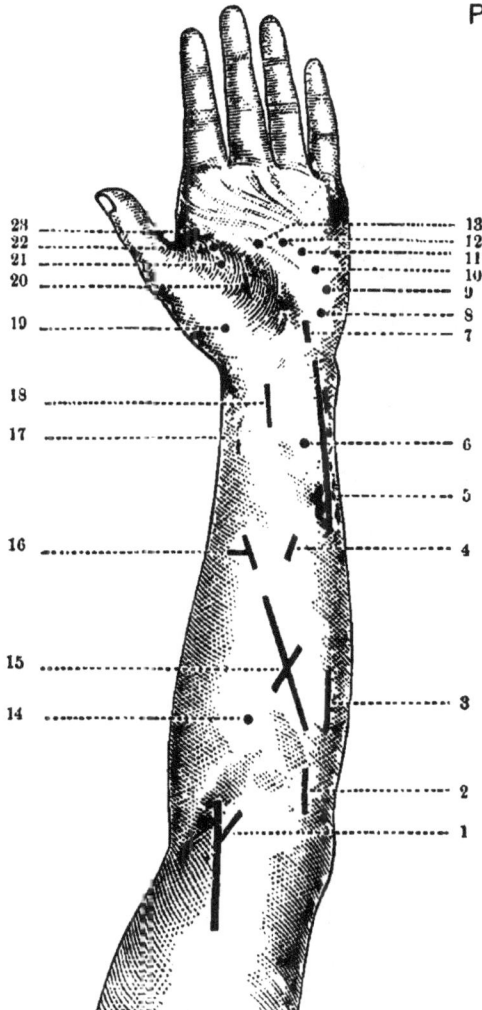

PLATE V.

THE MOTOR POINTS ON THE FLEXOR (ANTERIOR) ASPECT OF THE FOREARM.

1, median nerve and branch to m. pronator radii teres; 2, m. palmaris longus; 3, m.
flexor carpi ulnaris; 4, m. flexor sublim. digit.; 5, ulnar nerve; 6, m. flex.
sublim. dig.; 7, volar branch of the ulnar nerve; 8, m. palmaris brevis; 9, m.
abductor min. digit.; 10. m. flexor min. digit.; 11, m. opponens min. digit.;
12, 13, m. lumbricales; 14, m. flexor carpi radialis; 15, m. flexor profund.
digitorum; 16, m. flexor sublim. digitorum; 17, m. flex. longus pollicis; 18,
median nerve; 19, m. opponens pollicis; 20, m. abductor pollicis; 21, m. flexor
brevis pollicis; 22, m. adductor pollicis; 23, m. first lumbricalis.

PLATE VI.

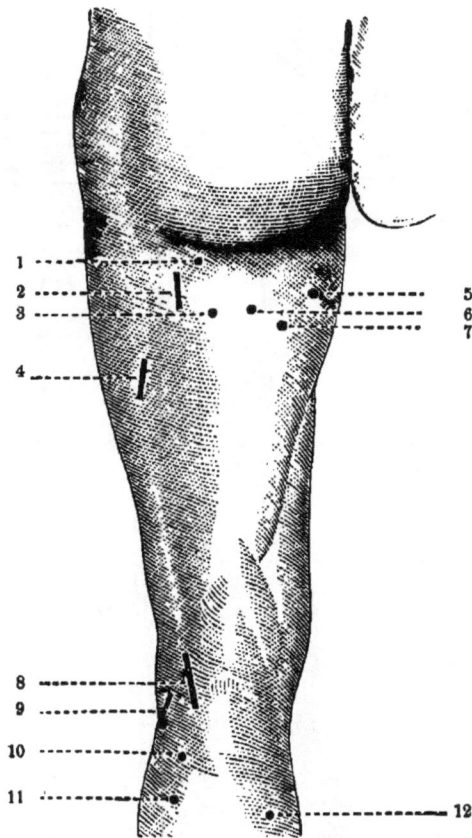

THE MOTOR POINTS ON THE POSTERIOR ASPECT OF THE THIGH.

1, branch of the inferior gluteal nerve to the gluteus maximus muscle; 2, sciatic nerve; 3, long head of biceps muscle; 4, short head of biceps muscle; 5, adductor magnus muscle; 6, semi-tendinosus muscle; 7, semi-membranosus muscle; 8, tibial nerve; 9, peroneal nerve; 10, external head of gastrocnemius muscle; 11, soleus muscle; 12, internal head of gastrocnemius muscle.

PLATE VII.

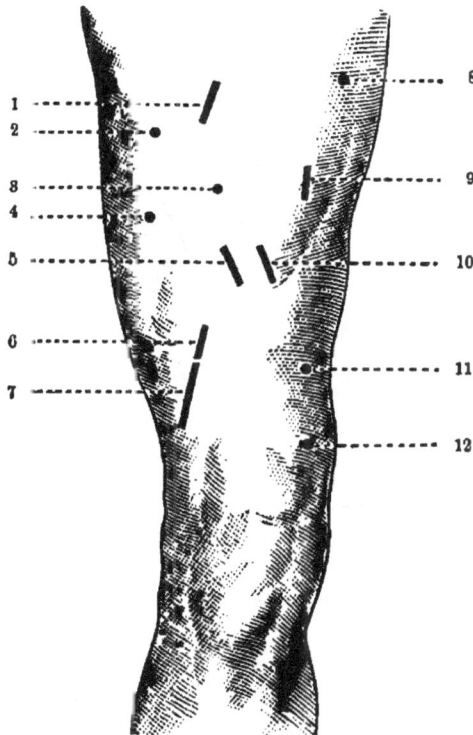

THE MOTOR POINTS ON THE ANTERIOR ASPECT OF THE THIGH.

1, crural nerve; 2, obturator nerve; 3, sartorius muscle; 4, adductor longus muscle; 5, branch of the anterior crural nerve for the quadriceps extensor muscle;
6, the quadriceps muscle; 7, branch of anterior crural nerve to the vastus inter-
nus muscle; 8, tensor vaginæ femoris muscle (supplied by the superior gluteal
nerve); 9, external cutaneous branch of anterior crural nerve; 10, rectus femoris muscle; 11, 12, vastus externus muscle.

PLATE VIII.

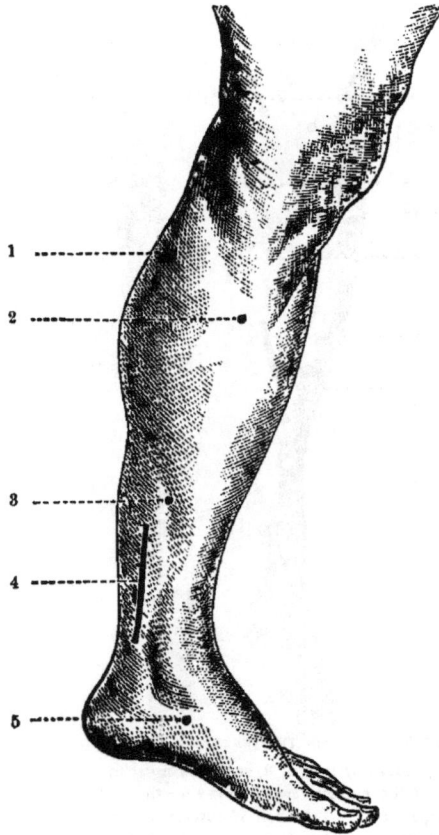

THE MOTOR POINTS ON THE INNER ASPECT OF THE LEG.

1, internal head of gastrocnemius muscle; 2, soleus muscle; 3, flexor communis digitorum muscle; 4, posterior tibial nerve; 5, abductor pollicis muscle.

PLATE IX.

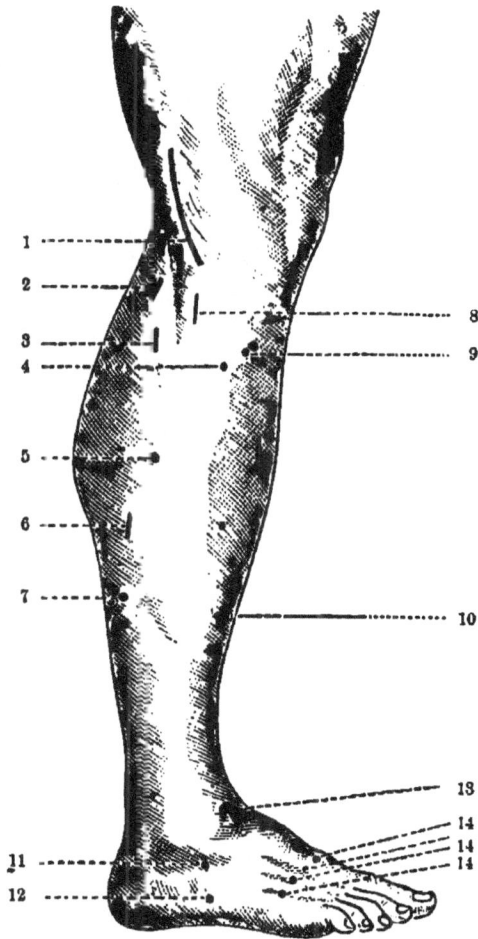

THE MOTOR POINTS ON THE OUTER ASPECT OF THE LEG.

1, peroneal nerve; 2, external head of gastrocnemius muscle; 3. soleus muscle; 4'
extensor communis digitorum muscle; 5, peroneus brevis muscle; 6, soleus
muscle; 7, flexor longus pollicis; 8. peroneus longus muscle; 9, tibialis anticus
muscle; 10, extensor longus pollicis muscle; 11, extensor brevis digitorum muscle; 12, abductor minimi digiti muscle; 13, deep branch of the peroneal nerve
to the extensor brevis digitorum muscle; 14, 14, 14, dorsal interossei muscles.

PLATE X.

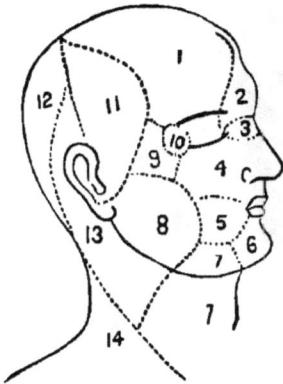

The Nervous Distribution of the Skin of the Head. (After Flower, but slightly modified.)

1, region supplied by the *supra-orbital* branch of the fifth nerve; 2, region supplied by the *supra-trochlear* branch of the fifth nerve; 3, region supplied by the *infra-trochlear* branch of the fifth nerve; 4, region supplied by the *infra-orbital* branch of the fifth nerve; 5, region supplied by the *buccal* branch of the fifth nerve; 6, region supplied by the *mental* branch of the fifth nerve; 7, region supplied by the *superficial cervical* from the cervical plexus; 8, region supplied by the *great auricular* from the cervical plexus; 9, region supplied by the *temporo-malar* branch of the fifth nerve; 10, region supplied by the *lachrymal* branch of the fifth nerve; 11, region supplied by the *auriculo-temporal* branch of the fifth nerve; 12, region supplied by the *great occipital* (a spinal nerve); 13, region supplied by the *small occipital* from the cervical plexus; 14, region supplied by the *supra-clavicular* from the cervical plexus.

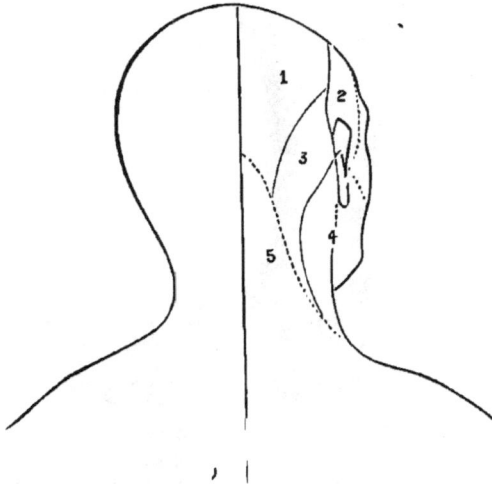

The Cutaneous Nerve Supply of the Posterior Portion of Head and Neck. (Modified from Flower.)

1, region supplied by the great occipital nerve; 2, region supplied by the auriculo-temporal nerve; 3, region supplied by the small occipital nerve; 4, region supplied by the great auricular nerve; 5, region supplied by the third cervical nerve.

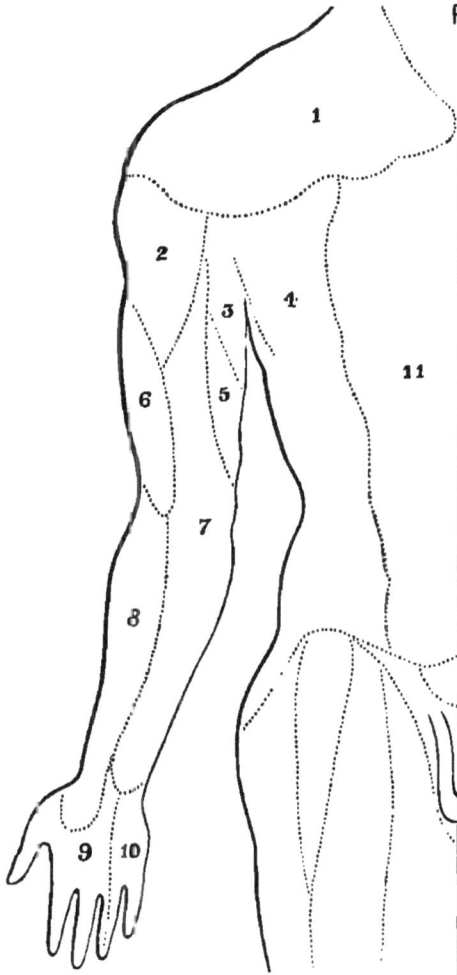

PLATE XI.

A DIAGRAM OF THE REGIONS OF CUTANEOUS NERVE DISTRIBUTION IN THE ANTERIOR SURFACE OF THE UPPER EXTREMITY AND TRUNK. (Modified from Flower.)

1, region supplied by the supra-clavicular nerve (branch of the cervical plexus); 2, region supplied by the circumflex nerve; 3, region supplied by the intercosto-humeral nerve; 4, region supplied by the intercostal nerve (lateral branch); 5, region supplied by the lesser internal cutaneous nerve (nerve of Wrisberg); 6, region supplied by the musculo-spiral nerve (external cutaneous branch); 7, region supplied by the internal cutaneous nerve; 8, region supplied by the musculo-cutaneous nerve; 9, region supplied by the median nerve; 10, region supplied by the ulnar nerve; 11, region supplied by the intercostal nerve (anterior branch).

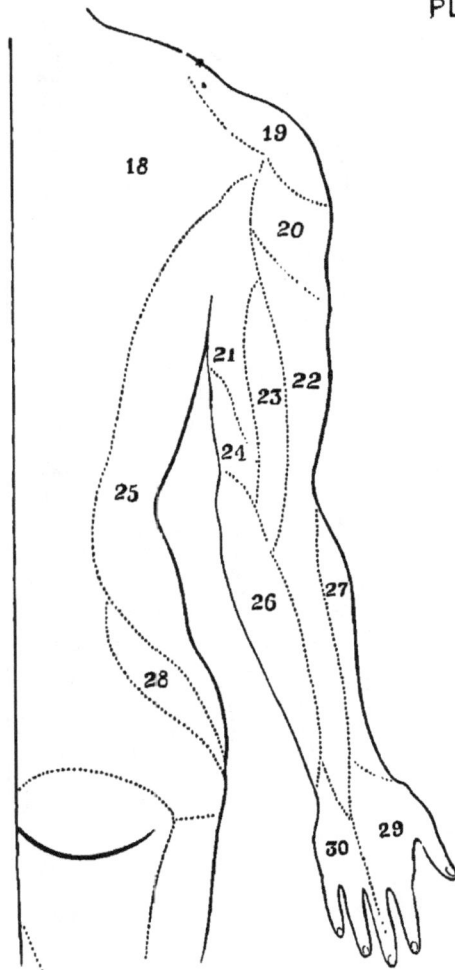

PLATE XII

A Diagram of the Regions of Cutaneous Nerve Distribution on the Posterior Surface of the Upper Extremity and Trunk. (Modified from Flower.)

18, region supplied by the *second dorsal* nerve; 19, region supplied by the *supra-
scapular* nerve; 20, region supplied by the *circumflex* nerve; 21, region supplied by the *intercosto-humeral* nerve; 22, region supplied by the *external
cutaneous* nerve; 23, region supplied by the *internal cutaneous branch* of the
musculo-spiral nerve; 24, region supplied by the "*nerve of Wrisberg*"; 25,
region supplied by the *lateral branches* of the *intercostal* nerves; 26, region
supplied by the *internal cutaneous* nerve; 27, region supplied by the *musculo-
cutaneous* nerve; 28, region supplied by the *iliac branch* of the *ilio-inguinal*
nerve; 29, region supplied by the *radial* nerve; 30, region supplied by the
ulnar nerve.

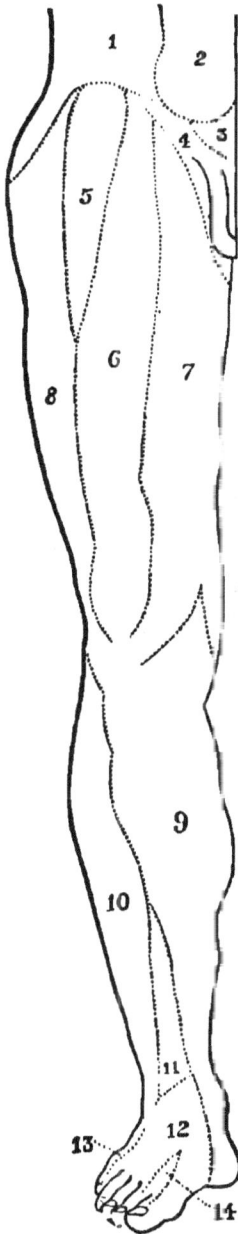

PLATE XIII.

A DIAGRAM OF THE CUTANEOUS SUPPLY OF THE ANTERIOR ASPECT OF THE LOWER EXTREMITY. (Modified from Flower.)

1, region supplied by the lateral branches of the intercostal nerves; 2, region supplied by the anterior branches of the intercostal nerves; 3, region supplied by the ilio-hypogastric nerve; 4, region supplied by the ilio-inguinal nerve; 5, region supplied by the genito-crural nerve; 6, region supplied by the middle cutaneous branch of the anterior crural nerve; 7, region supplied by the internal cutaneous branch of the anterior crural nerve and partly by the obturator nerve; 8, region supplied by the external cutaneous nerve; 9, region supplied by the long saphenous branch of the anterior crural nerve' 10, region supplied by the branches of the external popliteal nerve; 11, region supplied by the musculo-cutaneous nerve; 12, region supplied by the terminal filaments of the musculo-cutaneous nerve; 13, region supplied by the external saphenous nerve; 14, region supplied by the anterior tibial nerve.

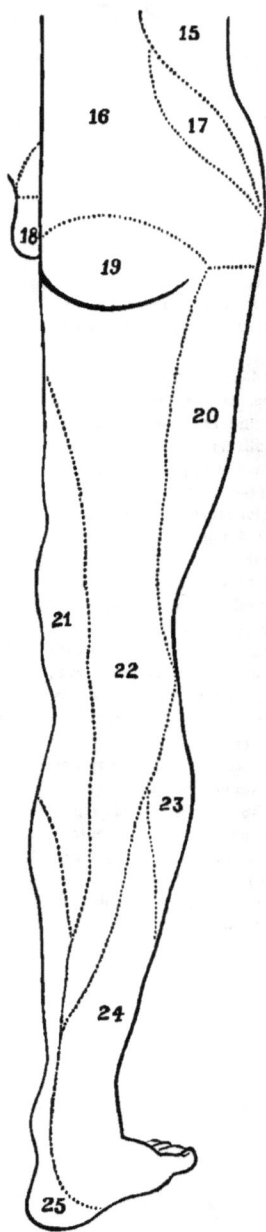

PLATE XIV.

A Diagram of the Cutaneous Supply
of the Posterior Aspect of the
Lower Extremities. (Modified from
Flower.)

15, region supplied by the lateral branches
of the intercostal nerves; 16, region
supplied by the posterior branches of
the lumbar nerves; 17, region supplied by the iliac branch of the ilio-
hypogastric nerve; 18, region supplied
by the pudic nerve; 19, region supplied by the inferior gluteal branch of
the small sciatic nerve; 20, region supplied by the external cutaneous nerve;
21, region supplied by the internal cutaneous branch of the anterior crural
nerve; 22, region supplied by the
small and great sciatic nerves; 23, region supplied by branches from the
external popliteal nerve; 24, region
supplied by the external saphenous
nerve; 25, region supplied by the
posterior tibial nerve.

THE

NEW YORK MEDICAL JOURNAL,

A WEEKLY REVIEW OF MEDICINE.

EDITED BY FRANK P. FOSTER, M. D.

THE LEADING JOURNAL OF AMERICA.

Containing twenty-eight double-columned pages of reading-matter, consisting of **Lectures, Original Communications, Clinical Reports, Correspondence, Book Notices, Leading Articles, Minor Paragraphs, News Items, Letters to the Editor, Proceedings of Societies, Reports on the Progress of Medicine, and Miscellany.**

By reason of the condensed form in which the matter is arranged, the JOURNAL contains more reading-matter than any other of its class in the United States. Its pages contain an average of 1,300 words; each volume has at least 748 pages, giving an aggregate of 972,400 words, or more than double the amount of reading-matter contained in a $5.00 octavo volume of 800 pages, averaging 500 words to the page. It is also more freely illustrated, and its illustrations are generally better executed, than is the case with other weekly journals.

The articles contributed to the JOURNAL are of a high order of excellence, for authors know that through its columns they address the better part of the profession; a consideration which has not escaped the notice of advertisers, as shown by its increasing advertising patronage.

The volumes begin with January and July of each year. Subscriptions can be arranged to begin with the volume.

TERMS, PAYABLE IN ADVANCE.

One Year - - - - - - - $5 00
Six Months - - - - - 2 50

The Popular Science Monthly and The New York Medical Journal to the same address, $9.00 per Annum (full price, $10.00).

New York: D. APPLETON & CO., 1, 3, & 5 Bond Street.

THE

POPULAR SCIENCE MONTHLY.

CONDUCTED BY E. L. AND W. J. YOUMANS.

THE POPULAR SCIENCE MONTHLY will continue, as heretofore, to supply its readers with the results of the latest investigation and the most valuable thought in the various departments of scientific inquiry.

Leaving the dry and technical details of science, which are of chief concern to specialists, to the journals devoted to them, the MONTHLY deals with those more general and practical subjects which are of the greatest interest and importance to the public at large. In this work it has achieved a foremost position, and is now the acknowledged organ of progressive scientific ideas in this country.

The wide range of its discussions includes, among other topics:

The bearing of science upon education;

Questions relating to the prevention of disease and the improvement of sanitary conditions;

Subjects of domestic and social economy, including the introduction of better ways of living, and improved applications in the arts of every kind;

The phenomena and laws of the larger social organizations, with the new standard of ethics, based on scientific principles;

The subjects of personal and household hygiene, medicine, and architecture, as exemplified in the adaptation of public buildings and private houses to the wants of those who use them;

Agriculture and the improvement of food-products;

The study of man, with what appears from time to time in the departments of anthropology and archæology that may throw light upon the development of the race from its primitive conditions.

Whatever of real advance is made in chemistry, geography, astronomy, physiology, psychology, botany, zoölogy, paleontology, geology, or such other department as may have been the field of research, is recorded monthly.

Special attention is also called to the biographies, with portraits, of representative scientific men, in which are recorded their most marked achievements in science, and the general bearing of their work indicated and its value estimated.

Terms: $5.00 per annum, in advance.

The New York Medical Journal and The Popular Science Monthly to the same address, $9.00 per annum (full price, $10.00).

New York: D. APPLETON & CO., 1, 3, & 5 Bond Street.

MEDICAL

AND

HYGIENIC WORKS

PUBLISHED BY

D. APPLETON & CO., 1, 3, & 5 Bond St., New York.

BARKER (FORDYCE). On Sea-Sickness. Small 12mo. Cloth, 75 cents.
On Puerperal Disease. Third edition. 8vo. Cloth, $5.00; sheep, $6.00.

BARTHOLOW (ROBERTS) A Treatise on Materia Medica and Therapeutics.
8vo. Cloth, $5.00; sheep, $6.00.

A Treatise on the Practice of Medicine. 8vo. Cloth, $5.00; sheep, $6.00.

On the Antagonism between Medicines and between Remedies and Diseases. 8vo.
Cloth, $1.25.

BASTIAN (H. CHARLTON). On Paralysis from Brain-Disease in its common
Forms. 12mo. Cloth, $1.75.

The Brain as an Organ of the Mind. 12mo. Cloth, $2.50.

BELLEVUE AND CHARITY HOSPITAL REPORTS. Edited by W. A. Hammond, M. D. 8vo. Cloth, $4.00.

BENNET (J. H.). Winter and Spring on the Shores of the Mediterranean. 12mo.
Cloth, $3.50.

On the Treatment of Pulmonary Consumption, by Hygiene, Climate, and Medicine. Thin 8vo. Cloth, $1.50.

BILLINGS (F. S.). The Relation of Animal Diseases to the Public Health, and their
Prevention. 8vo. Cloth $4.00.

BILLROTH (THEODOR). General Surgical Pathology and Therapeutics. 8vo.
Cloth. $5.00; sheep, $6.00.

BRAMWELL (BYROM). Diseases of the Heart and Thoracic Aorta. 8vo. Cloth,
$8.00; sheep, $9.00.

BUCK (GURDON), Contributions to Reparative Surgery. 8vo. Cloth, $3.00.

CARPENTER (W. B.). Principles of Mental Physiology, with their Application
to the Training and Discipline of the Mind, and the study of its Morbid Conditions. 12mo. Cloth, $3.00.

CARTER (ALFRED H.). Elements of Practical Medicine. 12mo. Cloth, $3.00.

CHAUVEAU (A.). The Comparative Anatomy of the Domesticated Animals. 8vo.
Cloth, $6.00.

COMBE (ANDREW). The Management of Infancy, Physiological and Moral. 12mo.
Cloth, $1.50.

COOLEY. Cyclopædia of Practical Receipts, and Collateral Information in the Arts,
Manufactures, Professions, and Trades, including Medicine, Pharmacy, and Domestic Economy. 2 vols., 8vo. Cloth, $9.00.

CORNING (J. L.). Brain Exhaustion, with some Preliminary Considerations on
Cerebral Dynamics. Crown 8vo. Cloth, $2.00.

DAVIS (HENRY G.). Conservative Surgery. 8vo. Cloth, $3.00.

ELLIOT (GEORGE T.). Obstetric Clinic. 8vo. Cloth, $4.50.

EVETSKY (ETIENNE). The Physiological and Therapeutical Action of Ergot.
8vo. Limp cloth, $1.00.

FLINT (AUSTIN). Medical Ethics and Etiquette. Commentaries on the National
Code of Ethics. 12mo. Cloth, 61 cents.

FLINT (AUSTIN, JR.). The Physiological Effects of Severe and Protracted Muscular Exercise; with Special Reference to its Influence upon the Excretion of
Nitrogen. 12mo. Cloth, $1.00.

Text-Book of Human Physiology. Imperial 8vo. Cloth, $6.00; sheep, $7.00.

FLINT (AUSTIN, JR.). The Source of Muscular Power. 12mo. Cloth, $1.00.

Manual Chemical Examinations of the Urine in Disease. 12mo. Cloth, $1.00.

FOURNIER (ALFRED). Syphilis and Marriage. 8vo. Cloth, $2.00; sheep, $3.00.

FREY (HEINRICH). The Histology and Histochemistry of Man. 8vo. Cloth, $5.00; sheep, $6.00.

FRIEDLANDER (CARL). The Use of the Microscope in Clinical and Pathological Examinations. 8vo. Cloth, $1.50.

GAMGEE (JOHN). Yellow Fever a Nautical Disease. 8vo. Cloth, $1.50.

GROSS (SAMUEL W.). A Practical Treatise on Tumors of the Mammary Gland. 8vo. Cloth, $2.50.

GUTMANN (EDWARD). The Watering-Places and Mineral Springs of Germany, Austria, and Switzerland. 12mo. Cloth, $2.50.

GYNÆCOLOGICAL TRANSACTIONS, VOL. VIII. Being the Proceedings of the Eighth Annual Meeting of the American Gynæcological Society, held in Philadelphia, September 18, 19, and 20, 1883. 8vo. Cloth. $5.00.

GYNÆCOLOGICAL TRANSACTIONS. VOL. IX Being the Proceedings of the Ninth Annual Meeting of the American Gynæcological Society, held in Chicago, September 30, and October 1 and 2, 1884. 8vo. Cloth, $5.00.

HAMILTON (ALLAN McL.). Clinical Electro-Therapeutics, Medical and Surgical. 8vo. Cloth, $2.00.

HAMMOND (W. A.). A Treatise on Diseases of the Nervous System. 8vo. Cloth, $5.00; sheep, $6.00.

A Treatise on Insanity, in its Medical Relations. 8vo. Cloth, $5.00; sheep, $6.00.

Clinical Lectures on Diseases of the Nervous System. 8vo. Cloth, $3.50.

HART (D. BERRY). Atlas of Female Pelvic Anatomy. Large 4to. (*Sold only by subscription.*) Cloth. $15.00.

HARVEY (A.). First Lines of Therapeutics. 12mo. Cloth, $1.50.

HEALTH PRIMERS. In square 16mo volumes. Cloth, 40 cents each.

I. Exercise and Training.	V. Personal Appearance in Health and Disease.
II. Alcohol: Its Use and Abuse.	
III. Premature Death: Its Promotion or Prevention.	VI. Baths and Bathing.
	VII. The Skin and its Troubles.
IV. The House and its Surroundings.	VIII. The Heart and its Functions.
IX. The Nervous System.	

HOFFMANN-ULTZMANN. Introduction to an Investigation of Urine, with Special Reference to Diseases of the Urinary Apparatus. 8vo. Cloth, $2.00.

HOWE (JOSEPH W.). Emergencies, and how to treat them. 8vo. Cloth, $2.50.

The Breath, and the Diseases which give it a Fetid Odor. 12mo. Cloth, $1.00.

HUXLEY (T. H.). The Anatomy of Vertebrated Animals. 12mo. Cloth, $2.50.

The Anatomy of Invertebrated Animals. 12mo. Cloth, $2.50.

JACCOUD (S.). The Curability and Treatment of Pulmonary Phthisis. 8vo. Cloth, $4.00.

JOHNSON (JAMES F. W.). The Chemistry of Common Life. 12mo. Cloth, $2.00.

JONES (H. MACNAUGHTON). Practical Manual of Diseases of Women and Uterine Therapeutics. 12mo. Cloth, $3.00.

KEYES (E. L.). The Tonic Treatment of Syphilis, including Local Treatment of Lesions. 8vo. Cloth, $1.00.

KINGSLEY (N. W.). A Treatise on Oral Deformities as a Branch of Mechanical Surgery. 8vo. Cloth, $5.00; sheep, $6.00.

LEGG (J. WICKHAM). On the Bile, Jaundice, and Bilious Diseases. 8vo. Cloth, $6.00; sheep, $7.00.

LETTERMANN (JONATHAN). Medical Recollections of the Army of the Potomac. 8vo. Cloth, $1.00.

LITTLE (W. J.). Medical and Surgical Aspects of In-knee (Genu-Valgum): Its Relation to Rickets, its Prevention, and its Treatment, with and without Surgical Operation. 8vo. Cloth, $2.00.

LORING (EDWARD G.). A Text-Book of Ophthalmoscopy. Part I. The Normal Eye, Determination of Refraction, and Diseases of the Media. 8vo. (*In press.*)

LUSK (WILLIAM T.). The Science and Art of Midwifery. Second edition, revised and enlarged. 8vo. Cloth, $5.00; sheep, $6.00.

LUYS (J.). The Brain and its Functions. 12mo. Cloth, $1.50.

McSHERRY (RICHARD). Health, and how to promote it. 12mo. Cloth, $1.25.

MARKOE (T. M.). A Treatise on Diseases of the Bones. 8vo. Cloth, $4.50.

MAUDSLEY (HENRY). Body and Mind: An Inquiry into their Connection and Mutual Influence, specially in reference to Mental Disorders. 12mo. Cloth, $1.50.

  Physiology of the Mind. 12mo. Cloth, $2.00.

  Pathology of the Mind. 12mo. Cloth, $2.00.

  Responsibility in Mental Disease. 12mo. Cloth, $1.50.

NEUMANN (ISIDOR). Hand-Book of Skin Diseases. 8vo. Cloth, $4.00; sheep, $5.00.

THE NEW YORK MEDICAL JOURNAL (weekly). Edited by Frank P. Foster, M. D. Terms per annum, $5.00.

  GENERAL INDEX, from April, 1865, to June, 1876 (23 volumes). 8vo. Cloth, 75 cents.

NIEMEYER (FELIX VON). A Text-Book of Practical Medicine, with particular reference to Physiology and Pathological Anatomy. 2 volumes, 8vo. Cloth, $9.00; sheep, $11.00.

NIGHTINGALE'S (FLORENCE) Notes on Nursing. 12mo. Cloth, 75 cents.

OSWALD (F. L.). Physical Education; or, The Health Laws of Nature. 12mo. Cloth, $1.00.

PEASLEE (E. R.). A Treatise on Ovarian Tumors: Their Pathology, Diagnosis, and Treatment, with reference especially to Ovariotomy. 8vo. Cloth, $5.00; sheep, $6.00.

PEREIRA'S (DR.) Elements of Materia Medica and Therapeutics. Royal 8vo. Cloth, $7.00; sheep, $8.00.

PEYER (ALEX.). Clinical Microscopy. Translated by A. C. Girard, M. D. (*Nearly ready.*)

POMEROY (OREN D.). The Diagnosis and Treatment of Diseases of the Ear. 8vo. $3.00.

POORE (C. T.). Osteotomy and Osteoclasis, for the Correction of Deformities of the Lower Limbs. 8vo. Cloth, $2.50.

QUAIN (RICHARD). A Dictionary of Medicine, including General Pathology, General Therapeutics, Hygiene, and the Diseases peculiar to Women and Children. By Various Writers. Edited by Richard Quain, M. D. 8vo. (*Sold only by subscription.*) Half morocco, $8.00.

RANNEY (AMBROSE L.). Applied Anatomy of the Nervous System. 8vo. Cloth, $4.00; sheep, $5.00.

RIBOT (TH.). Diseases of Memory: An Essay in the Positive Psychology. 12mo. Cloth, $1.50.

RICHARDSON (B. W.). Diseases of Modern Life. 12mo. Cloth, $2.00.

  A Ministry of Health and other Addresses. 12mo. Cloth, $1.50.

ROBINSON (A. R.). A Manual of Dermatology. 8vo. Cloth, $5.00.

ROSENTHAL (I.). General Physiology of Muscles and Nerves. 12mo. Cloth, $1.50.

ROSCOE AND SCHORLEMMER. Treatise on Chemistry.
- Vol. 1. Non-Metallic Elements. 8vo. Cloth, $5.00.
- Vol. 2. Part 1. Metals. 8vo. Cloth, $3.00.
- Vol. 2. Part II. Metals. 8vo. Cloth, $3.00.
- Vol. 3. Part I. The Chemistry of the Hydrocarbons and their Derivatives. 8vo. Cloth, $5 00.
- Vol. 3. Part II. The Chemistry of the Hydrocarbons and their Derivatives. 8vo. Cloth, $5.00.

SAYRE (LEWIS A.). Practical Manual of the Treatment of Club-Foot. 12mo. Cloth, $1.25.

Lectures on Orthopedic Surgery and Diseases of the Joints. 8vo. Cloth, $5.00; sheep, $6.00.

SCHROEDER (KARL). A Manual of Midwifery, including the Pathology of Pregnancy and the Puerperal State. 8vo. Cloth, $3.50; sheep, $4.50.

SIMPSON (JAMES Y.). Selected Works: Anæsthesia, Diseases of Women. 3 volumes, 8vo. Per volume, cloth, $3.00; sheep, $4.00.

SMITH (EDWARD). Foods. 12mo. Cloth, $1.75.

Health. 12mo. Cloth, $1.00.

STEINER (JOHANNES). Compendium of Children's Diseases. 8vo. Cloth, $3.50; sheep, $4.50.

STONE (R. FRENCH). Elements of Modern Medicine, including Principles of Pathology and of Therapeutics, with many Useful Memoranda and Valuable Tables of Reference. Accompanied by Pocket Fever Charts. Wallet-book form, with pockets on each cover for Memoranda, Temperature Charts, etc. $2.50.

STRECKER (ADOLPH). Short Text-Book of Organic Chemistry. 8vo. Cloth, $5.00.

SWANZY (HENRY R.). A Hand-Book of the Diseases of the Eye, and their Treatment. Crown 8vo. Cloth, $3.00.

TRACY (ROGER S.). The Essentials of Anatomy, Physiology, and Hygiene. 12mo. Cloth, $1.25.

Hand-Book of Sanitary Information for Householders. 16mo. Cloth, 50 cents.

TRANSACTIONS OF THE NEW YORK STATE MEDICAL ASSOCIATION, VOLUME I. Being the Proceedings of the First Annual Meeting of the New York State Medical Association, held in New York, November 18, 19, and 20, 1884. Small 8vo. Cloth, $5.00.

TYNDALL (JOHN). Essays on the Floating Matter of the Air, in Relation to Putrefaction and Infection. 12mo. Cloth, $1.50.

ULTZMANN (ROBERT). Pyuria, or Pus in the Urine, and its Treatment. 12mo. Cloth, $1.00.

VAN BUREN (W. H.). Lectures upon Diseases of the Rectum, and the Surgery of the Lower Bowel. 8vo. Cloth, $3.00; sheep, $4.00.

Lectures on the Principles and Practice of Surgery. 8vo. Cloth, $4.00; sheep, $5.00.

VAN BUREN AND KEYES. A Practical Treatise on the Surgical Diseases of the Genito-Urinary Organs, including Syphilis. 8vo. Cloth, $5.00 ; sheep, $6.00.

VOGEL (A.). A Practical Treatise on the Diseases of Children. Third American from the eighth German edition, revised and enlarged. 8vo. Cloth, $4.50; sheep, $5.50.

WAGNER (RUDOLF). Hand-Book of Chemical Technology. 8vo. Cloth, $5 00.

WALTON (GEORGE E.). Mineral Springs of the United States and Canadas. 12mo. Cloth, $2.00.

WEBBER (S. G.). A Treatise on Nervous Diseases. 8vo. Cloth, $2.50.

WELLS (T. SPENCER). Diseases of the Ovaries. 8vo. Cloth, $4.50.

WYLIE (WILLIAM G.). Hospitals: Their History, Organization, and Construction. 8vo. Cloth, $2.50.

www.ingramcontent.com/pod-product-compliance
Lightning Source LLC
Chambersburg PA
CBHW020538270326
41927CB00006B/628